T0302235

Probing Mechanics at
Nanoscale Dimensions

MATERIALS RESEARCH SOCIETY
SYMPOSIUM PROCEEDINGS VOLUME 1185

Probing Mechanics at Nanoscale Dimensions

Symposium held April 14–17, 2009, San Francisco, California, U.S.A.

EDITORS:

Nobumichi Tamura
Lawrence Berkeley National Laboratory
Berkeley, California, U.S.A.

Andrew Minor
University of California, Berkeley and
Lawrence Berkeley National Laboratory
Berkeley, California, U.S.A.

Conal Murray
IBM T.J. Watson Research Center
Yorktown Heights, New York, U.S.A.

Lawrence Friedman
The Pennsylvania State University
University Park, Pennsylvania, U.S.A.

Materials Research Society
Warrendale, Pennsylvania

CAMBRIDGE
UNIVERSITY PRESS

University Printing House, Cambridge CB2 8BS, United Kingdom

One Liberty Plaza, 20th Floor, New York, NY 10006, USA

477 Williamstown Road, Port Melbourne, VIC 3207, Australia

314-321, 3rd Floor, Plot 3, Splendor Forum, Jasola District Centre, New Delhi - 110025, India

79 Anson Road, #06-04/06, Singapore 079906

Cambridge University Press is part of the University of Cambridge.

It furthers the University's mission by disseminating knowledge in the pursuit of education, learning and research at the highest international levels of excellence.

www.cambridge.org
Information on this title: www.cambridge.org/9781605111582

Materials Research Society
506 Keystone Drive, Warrendale, PA 15086
http://www.mrs.org

© Materials Research Society 2009

First published 2009
First paperback edition 2012

Single article reprints from this publication are available through University Microfilms Inc., 300 North Zeeb Road, Ann Arbor, MI 48106

CODEN: MRSPDH

A catalogue record for this publication is available from the British Library

ISBN 978-1-605-11158-2 Hardback
ISBN 978-1-107-40819-7 Paperback

Effort sponsored by the Air Force Office of Scientific Research, Air Force Material Command, USAF, under FA9550-09-1-0541. The U.S. Government is authorized to reproduce and distribute reprints for Governmental purposes notwithstanding any copyright notation thereon. The views and conclusions herein are those of the authors and should not be interpreted as necessarily representing the official policies or endorsements, either expressed or implied, of the Air Force Office of Scientific Research or the U.S. Government.

*Invited Paper

PREFACE

These proceedings are a record of Symposium II, "Probing Mechanics at Nanoscale Dimensions," held April 14–17 at the 2009 MRS Spring Meeting in San Francisco, California.

Mechanical properties and the reliability of materials greatly depend on the details of their microstructure. However, most engineered materials, which are often polycrystalline and multiphase in nature and have undergone a number of processing steps, are extremely complex and inhomogeneous at the local level. The precise relationship between microstructure and physical properties for these types of materials is an issue that becomes even more critical as device dimensions rapidly decrease toward nanoscale dimensions (nanomaterials and NEMS).

During the last decade, new experimental tools have emerged, allowing us to access information on the microstructure and state of deformation of materials at a fine spatial resolution ranging from microns down to tens of nanometers. In parallel, developments in computational materials simulation are now able to incorporate discretization (grain, grain boundaries, and defects) into modeling, which is a necessary step to obtain a thorough multiscale, theoretical understanding of material properties. This symposium was aimed to cover both the theoretical and experimental aspects on how to define and measure stress, strain, and the deformation of materials at the appropriate microstructural level of grain, grain boundaries, and other defects.

The order of the papers in this volume follows the order of their presentation at the MRS Meeting. Papers presented during posters sessions are at the end of the proceedings.

Nobumichi Tamura
Andrew Minor
Conal Murray
Lawrence Friedman

June 2009

ACKNOWLEDGMENTS

We would like to thank all the people who contributed to make this symposium a success: the speakers and poster presenters for their outstanding presentations at the meeting, the authors who devoted time and energy to further contribute to the present proceedings volume, the session chairs who did a wonderful job in making sure the symposium was held in a timely manner, the reviewers who made sure that the technical papers included in these proceedings are of outstanding quality, the Materials Research Society staff and Meeting Chairs who made sure that the organization of a symposium was a pleasant experience for us, the symposium assistants for the technical support provided during the meeting and last but not least, the following sponsors for their generous financial contribution:

Air Force Office of Scientific Research
Hysitron Inc
IBM T.J. Watson Research Center

MATERIALS RESEARCH SOCIETY SYMPOSIUM PROCEEDINGS

MATERIALS RESEARCH SOCIETY SYMPOSIUM PROCEEDINGS

Mater. Res. Soc. Symp. Proc. Vol. 1185 © 2009 Materials Research Society   1185-II02-07

Effect of oxygen on nanoscale indentation-induced phase transformations in amorphous silicon

S. Ruffell[1] and J. S. Williams[1]

[1]Department of Electronic Materials Engineering, Research School of Physics and Engineering, Australian National University, Canberra, ACT, 0200, Australia

ABSTRACT

Ion-implantation has been used to introduce oxygen concentration-depth profiles into nominally oxygen-free amorphous silicon (a-Si). The effect of O concentrations in excess of 10^{18} cm^{-3} on the formation of high pressure crystalline phases (Si-III and Si-XII) during indentation unloading has been studied. By examination of unloading curves and post-indent Raman micro-spectroscopy O is found to inhibit the so-called pop-out event during unloading and, therefore, the formation of the crystalline phases. Furthermore, at high O concentrations (> 10^{21} cm^{-3}) the formation of these phases is reduced significantly such that under indentation conditions used here the probability of forming the phases is reduced to almost zero. We suggest that the bonding of O with Si reduces the formation of Si-III/XII during unloading through a similar mechanism to that of oxygen-retarded solid phase crystallization of a-Si.

INTRODUCTION

Nanoindentation-induced phase transformations in Si have attracted significant interest over the last few decades with more recent studies reported in nanoindentation of ion-implanted amorphous Si (a-Si) [2-6]. During loading, a transformation to the β-Sn phase (Si-II) occurs at a critical pressure of ~12 GPa. On unloading, the Si-II further transforms to either amorphous silicon (a-Si) or a mixture of high pressure polycrystalline phases (Si-III and Si-XII); the latter being favoured for slow unloading and is usually accompanied by a pop-out event [2, 3]. These pressure-induced transformations are well characterized [2-6], but the exact mechanisms behind the phenomena are still not well understood. In particular the proposed formation of the crystalline phases by a nucleation and growth mechanism has not been studied in detail. Recent studies have investigated these phase transformations during indentation of ion-implanted a-Si [5, 7]. It was found that the Si-III/Si-XII phases form more readily in an a-Si matrix compared to c-Si e.g. volumes of Si-III/Si-XII formed in a-Si with unloading rates over 3 orders of magnitude greater than the unload rates required to form the phases in c-Si. However, recent work by the current authors on plasma enhanced chemical vapour deposited (PECVD) a-Si films found that the films do not undergo these phase transformations. The reasons for this are not understood but one possibility is that the high impurity content in the deposited films compared to a "pure" film created by ion-implantation prevents the formation of the high pressure phases. In particular, O and H are found in high concentrations (10^{19} to 10^{21} cm^{-3}) in PECVD deposited films compared to $\leq10^{18}$ cm^{-3} in a-Si formed by Si ion-implantation [8]. The aim of this study is to study the effect of O on the indentation-induced phase transformations. This is done through ion-implantation of O in to ion-implanted a-Si, controllably adding a range of O concentrations into a "pure" a-Si layer over the depth range of the phase transformed zones formed by subsequent indentation. Indentation is performed under conditions that ensure a high probability of forming Si-III/XII in the phase transformed zones for nominally O-free ion-implanted a-Si. Analysis of the load/unload curves and Raman micro-spectroscopy are used to study the effect of O on the phase transformation behaviour.

EXPERIMENT

All samples were fabricated in a 7 μm epi-layer p-doped with boron to a resistivity of 10-20 Ω-cm grown on a low resistivity (0.002 Ω-cm) Si(100) wafer. A 350 nm thick surface layer of a-Si was created by multiple energy implantation of Si. Samples were then cleaved and implanted with O. Implantation of O was performed at 30 and 50 keV to total fluences of 3.1×10^{14}, 3.1×10^{15}, and 3.1×10^{16} cm^{-2} corresponding to peak O concentrations of $\sim 2 \times 10^{19}$, 2×10^{20}, and 2×10^{21} cm^{-3} (fig. 1). These concentrations cover those typically found in deposited a-Si samples made by various a-Si deposition methods. The concentration-depth profiles for all oxygen implanted samples following a relaxation anneal of 450 °C for 30 minutes [9] are shown in figure 1. The final set of samples consisted of a sample containing no additionally implanted O (labeled a-Si) and O implanted samples which are referred to by the total implanted fluence.

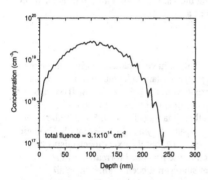

Figure 1. Simulation of the O implant profile [1]. The figure shows the concentration-depth profile for the implanted fluence of 3.1×10^{14} cm^{-2}. The other implant fluences scale with this profile.

Indentation was performed using a Hysitron Triboindenter fitted with a Berkovich diamond tip. Loading to 4 mN (and 7 mN) and unloading at 0.2 mN/s was repeated 54 times in each sample. These conditions form a phase transformed zone that extends 150 nm below the surface and has a diameter of ~400 nm for the 4 mN maximum load (the zone extends approximately 250 nm below the surface for 7 mN). These zones have been imaged using XTEM (not shown here). Typical load/unload curves and indentation hardness data are shown in figure 2 for the samples. These loading conditions result in a probability of pop-out during unloading of ~0.4 for nominally O-free a-Si of thickness 350 nm (approx. 0.8 for 7mN). A prime indicator for the formation of these phases during unloading is the presence of a pop-out on the unloading curve. The formation of Si-III/XII proceeds through a nucleation and growth process which results in a variation in final microstructure between indents made under identical conditions. Therefore, the probability of a pop-out occurring during unloading and the formation of Si-III/XII was extracted from both analysis of the load/unload curves and Raman spectra from the series of 54 indents made in each sample.

Following the indentation tests, every residual indent was measured by Raman spectroscopy using a Renishaw 2000 instrument fitted with a HeNe laser focused to a spot of ~1 μm diameter (power 2.1mW). These measurements provide a method for detecting the presence of Si-III/XII. The measurements can also be correlated with the load/unload curves from the indentation tests.

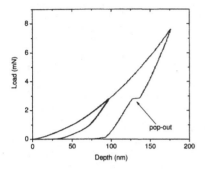 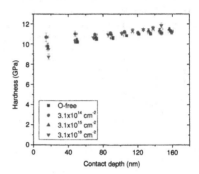

Figure 2. (left) Typical load/unload curves for indentation in all samples. A pop-out is observable for this example for the 7mN indent. (right) Indentation hardness versus depth data for all samples extracted from a series of indents with increasing load up to ~10 mN. No appreciable difference in mechanical properties is observed across samples.

RESULTS AND DISCUSSION

Figure 3 shows the probability of a pop-out occurring as a function of implanted O fluence for loading to 4 mN and unloading at a rate of 0.2 mN/s. The probability of a pop-out occurring decreases with increasing O content. For a fluence of 3.1×10^{16} cm^{-2} (corresponding to a peak O concentration of ~2×10^{21} cm^{-3}) the probability is reduced to only ~0.03 which is comparable to that of c-Si (also shown in fig. 3).

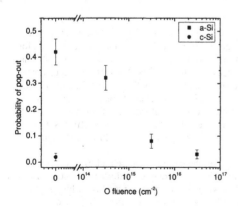

Figure 3. Probability of observing a pop-out event during unloading as a function of total implanted O fluence. Unloading from 4 mN at a rate of 0.2 mN/s was performed and the data were extracted from 54 indents made in each sample. Also, shown is the probability of a pop-out for indentation in c-Si.

Figure 4 shows typical Raman spectra taken from indents made in nominally O-free and O-implanted a-Si samples at two different maximum loads (4 and 7 mN). For the O-free a-Si, extra peaks associated with Si-III/Si-XII are visible around 350-420 cm^{-1}. The peaks are more intense for the larger indents due to the larger volumes of Si-III/Si-XII formed in these indents. For a maximum load of 4 mN, these peaks are not visible in the spectra taken from all fluences of O-implanted samples. This means that any Si-III/Si-XII in the residual indents is formed at volumes below the detection limit for Raman. For the larger indents, Si-III/Si-XII is visible in the spectra for the 3.1x10^{14} and 3.1x10^{15} cm^{-2} samples. However, the volume clearly decreases with increasing O until no Si-III/Si-XII is detected for the 3.1x10^{16} cm^{-2} sample.

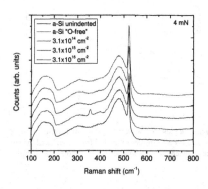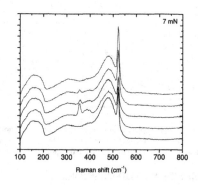

Figure 4. Typical Raman spectra taken in unindented a-Si and indents in samples with increasing O concentration. The left panel is for data taken in 4 mN indents. The right panel shows data for 7 mN indents. The small peaks between 300 and 400 cm^{-1} are associated with Si-III and Si-XII.

Figure 5 summarizes the Raman data taken on the indents made at 4 and 7 mN. The probability of detecting Si-III/XII in the indents decreases from 1 to 0 at a O fluence of 3.1x10^{14} cm^{-2} for loading to 4 mN. For the larger indents made at 7 mN, this decrease occurs between 3.1x10^{15} and 3.1x10^{16} cm^{-2}.

Although not all of the O-free indents exhibit a clear pop-out event (fig. 2 and 3), Si-III/Si-XII was detected in all indents made in the O-free sample (see fig. 5). It has been shown previously that Si-III/XII can still be formed in the absence of a clear pop-out e.g. for indentation at elevated temperatures or when only a small fraction of the phase transformed zone transforms to Si-III/XII [10, 11]. It was suggested that in these cases, the Si-III/Si-XII was formed in a more continuous fashion than a sudden catastrophic formation of a substantial volume. In addition, the current authors have never observed the absence of Si-III/Si-XII (from Raman and transmission electron microscopy measurements) when a pop-out occurs on unloading. Although Si-III/Si-XII is not detected by Raman in many of these implanted samples, it is likely that Si-III/Si-XII is still formed in the indents that exhibit a pop-out on unloading but at a volume below the Raman detection limit.

For many load/unload curves (not shown here) the pop-out is more kink-like as the O content increases. The pop-out event is associated with the sudden formation of a substantial volume of Si-III/Si-XII. The density of these phases is lower than that of the Si-II which forms under the indenter tip during loading. Thus, the transformation of a significant volume of Si-II to Si-III/XII results in a volume increase beneath tip and forces it out from the surface. The magnitude of the pop-out event will thus decrease as the volume of Si-III/Si-XII formed decreases.

The larger volume and more ready detection of Si-III/XII in the 7 mN indents is likely a result of two factors. One, the volume of Si-II formed during loading is larger which promotes the formation of Si-III/XII; two, the phase transformed zone extends to a depth where the O concentrations decrease to the background level. In this region the formation of Si-III/XII is inhibited to a lesser degree than at the peak of the O concentration profile (at a depth of ~100 nm).

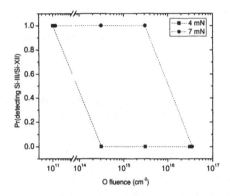

Figure 5. Summary of Raman data. Probability of detecting Si-III/Si-XII is shown as a function of implanted O fluence for indents made with a maximum load of 4 and 7 mN.

When the a-Si transforms to a metallic phase during loading it is unclear whether this phase is Si-II or a high density a-Si phase [12, 13]. However, this metallic phase does transform to Si-III/XII during unloading [5, 7, 12, 14]. We note that the Si-III and Si-XII phases only form from such a metallic phase during unloading. Therefore, it could be possible when no Si-III/Si-XII is observed at high O concentrations, that the reduced high pressure phase formation is due to O inhibiting the transformation from a-Si to the metallic phase during loading. However, this seems unlikely as no change in mechanical response of the Si is observed in the load/unload data, whereas it was shown previously that when unrelaxed ion-implanted a-Si doesn't transform during loading, the material is significantly softer [14]. The measured hardness versus contact depth for a series of indents made in nominally O-free a-Si and O implanted samples is shown in fig 2. These data indicate then that the metallic phase does form during loading (and is largely unaffected by oxygen) whereas the formation of Si-III/XII is effected by the O during unloading. Additionally, when reduced volumes of Si-III/XII are measured the remainder of the transformed zone, which is composed of a-Si is also formed through phase transformation from a metallic Si phase.

The presence of high concentrations of O (~0.5 at. %) in a-Si has been found previously to retard thermally induced solid phase crystallization by an order of magnitude [15]. Strong Si-O bonds, such as those that will be formed here, effect the kinetics of bond-breaking required for rearrangement of atoms in the a-Si to the crystalline form [15, 16]. A similar mechanism can be envisaged here to inhibit the bond rearrangement required for the transformation of the metallic phase to the high pressure crystalline phases, Si-III and Si-XII, noting that the highest concentration of O in this study (10^{21} cm^{-3}) corresponds to 2 at. %.

CONCLUSIONS

The effect of local O concentration on the nanoindentation-induced phase transformations in ion-implanted a-Si has been studied. Ion-implantation of O has been used to create concentration-depth profiles extending to ~200 nm below the surface with peak concentrations of up to 2×10^{21} cm^{-3}. The probability of the occurrence of a pop-out, indicating the formation of Si-III/Si-XII during unloading, decreases with increasing O content. Raman measurements made on residual indents confirm that the volume of Si-III/Si-XII formed decreases with oxygen content. We suggest that the bonding of O with Si reduces the formation of Si-III/XII during unloading through a similar mechanism to that of oxygen-retarded solid phase crystallization of a-Si.

ACKNOWLEDGMENTS

The authors gratefully acknowledge funding from the Australian Research Council and WRiota Pty Ltd.

REFERENCES

1. J. F. Ziegler, J. P. Biersack, *The stopping and range of ions in matter (SRIM)*. 2003.
2. J. E. Bradby, J. S. Williams, J. Wong-Leung, M. V. Swain, P. Munroe, *Journal of Materials Research* **16**, 1500 (2001).
3. V. Domnich, Y. Gogotsi, S. Dub, *Applied Physics Letters* **76**(16), 2214 (2000).
4. G. M. Pharr, W. C. Oliver, R. F. Cook, P. D. Kirchner, M. C. Kroll, *et al.*, *Journal of Materials Research* **7**, 961 (1992).
5. S. Ruffell, J. E. Bradby, J. S. Williams, *Applied Physics Letters* **89**(9), 091919 (2006).
6. J.S. Williams, Y. Chen, J. Wong-Leung, A. Kerr, M.V. Swain, *Journal of Materials Research* **14**, 2338 (1999).
7. S. Ruffell, J. E. Bradby, J. S. Williams, P. Munroe, *Journal of Applied Physics* **102**, 063521 (2007).
8. H. M. Liaw, *Semiconductor Instruments* **2**, 71 (1979).
9. *This anneal converts the a-Si to a relaxed state which is required for subsequent phase transformation during indentation.*
10. S. Ruffell, J. Vedi, J. E. Bradby, J. S. Williams, B. Haberl, *Journal of Applied Physics* **105**, 083520 (2009).
11. S. Ruffell, J. E. Bradby, J. S. Williams, D. Munoz-Paniagua, S. Tadayyon, *et al.*, *Nanotechnology* **20**, 135603 (2009).
12. Sudip K. Deb, Martin Wilding, Maddury Somayazulu, Paul F. McMillan, *Letters to Nature* **414**, 528 (2001).
13. O. Shimomura, S. Minomura, N. Sakai, K. Asaumi, K. Tamura, *et al.*, *Philosophical Magazine* **29**, 547 (1974).
14. B Haberl, J.E. Bradby, S. Ruffell, J.S. Williams, P Munroe, *Journal of Applied Physics* **100**, 013520 (2006).
15. J. S. Williams, R. G. Elliman, *Physical Review Letters* **51**(12), 1069 (1983).
16. E. F. Kennedy, L. Csepregi, J. W. Mayer, T. W. Sigmon, *Journal of Applied Physics* **48**(10), 4241 (1977).

Mater. Res. Soc. Symp. Proc. Vol. 1185 © 2009 Materials Research Society   1185-II02-08

Measuring Local Mechanical Properties Using FIB Machined Microcantilevers

David E. J. Armstrong[1], Angus J. Wilkinson[1], and Steve G. Roberts[1]
[1]Department of Materials Science, University of Oxford, Parks Road
Oxford, OX1 3PH, United Kingdom

ABSTRACT

Micro-scale Focused Ion Beam (FIB) machined cantilevers were manufactured in single crystal copper, polycrystalline copper and a copper-bismuth alloy. These were imaged and tested in bending using a nanoindenter. Cantilevers machined inside a single grain of polycrystalline copper were tested to determine their (anisotropic) Young's modulus: results were in good agreement with values calculated from literature values for single crystal elastic constants. The size dependence of yield behavior in the Cu microcantilevers was also investigated. As the thickness of the specimen was reduced from 23μm to 1.7μm the yield stress increased from 300MPa to 900MPa. Microcantilevers in Cu-0.02wt%Bi were manufactured containing a single grain boundary of known character, with a FIB-machined sharp notch on the grain boundary. The cantilevers were loaded to fracture allowing the fracture toughness of grain boundaries of different misorientations to be determined.

INTRODUCTION

Much recent work on measuring mechanical properties on the microscale has been carried out using focused ion beam (FIB) machining to manufacture a wide range of different tests specimens[1-3] which are then tested using a nanoindenter as a loading device. This is of great interest as it allows mechanical properties to be measured from much smaller volumes than are required for traditional mechanical tests. It is now possible to measure directly the mechanical properties of individual microstructural features such as grains, grain boundaries or thin layers, such as ion-implanted layers in bulk specimens [4].

A recent paper [5] by the present authors has demonstrated that testing of microcantilevers with a triangular cross section, machined into the surface of single crystal copper, can be used to measure the anisotropy of Young's modulus with respect to crystallography. This paper describes the extension of such techniques to measure plastic and fracture properties as well as Young's modulus.

EXPERIMENT

Materials

Three materials were chosen as model systems for developing microcantilever-testing techniques. For measuring elastic properties high purity polycrystalline copper was chosen. This was easily prepared and is well known for having highly anisotropic elastic properties. Single crystal copper was used for carrying out measurements into size effects in plastic properties. This allowed all cantilevers to be made with the same crystallography to avoid the influence of crystal anisotropy. For measuring grain-boundary fracture properties Cu-Bi was chosen, as it is well known that a small amount of bismuth (less then 0. 1wt%) [6] causes grain boundary embrittlement at room temperature.

Bulk Sample Preparation

The single crystal copper (99.9999%) was prepared by first mechanically polishing with diamond pastes of decreasing size from 8μm to 1μm. This was followed by electropolishing using 1.2V and 0.1A for 30 seconds in a solution of 80% ortho-phosphoric acid, producing a smooth surface free from mechanical deformation. EBSD was used to confirm the surface normal as [110] and to indentify the in-plane [001] and [1 $\overline{1}$ 0] directions.

Polycrystalline copper bar (99.9999%) was annealed at 800°C for 1 hour in argon to produce a large-grain sample. It was then sectioned to 500μm thick discs, which were ground on SiC paper of 1200 grit, followed by vibropolishing on colloidal silica for 24 hours. EBSD was used to determine the average grain size (~200μm) and to confirm that the surface was free from mechanical deformation.

To manufacture the copper-bismuth alloy, 9.7g of Cu and 3.7mg of Bi were sealed together in an evacuated clean quartz tube. This was heated in an argon atmosphere at 1100°C for 1 hour and then slow-cooled to room temperature. This produced an ingot of 4mm diameter, which was swaged to reduce the diameter to 3mm, producing an elongated grain structure with most grain boundaries running along the length of the ingot. The ingot was sectioned using a slow saw and polished using SiC paper to 1200 grit and then colloidal silica on a vibropolishing wheel for 24hours. This was used as electropolishing has been shown to alter boundary chemistry in copper-bismuth alloys [7].The sample was kept in the cold worked stat to ensure boundaries were very close to normal of the surface.

FIB Machining Of Microcantilevers

Each microcantilever was manufactured using a FEI FIB200. For the studies into elastic and plastic properties the same beam design was used: a cantilever of triangular cross section machined into the surface of a bulk sample. Large beam currents, typically 3000pA to 5000pA, were used for the initial steps. First a "U" shaped trench was milled. The sample was then tilted to 30° and undercut from each side. Smaller beam currents, 300pA to 1000pA, were then used to clean each side of the cantilever to remove any redeposited material. For measuring elastic properties of copper a standard cantilever size was used, with width 3μm, thickness 3.5μm and length 27μm. These were placed inside 4 different grains, which were identified using EBSD, Figure 1 d. This then allowed the crystallographic direction of the long angle of the cantilever to be found. Size effects in yield stress were measured using cantilevers of varying size ranging from 1μm wide, 1.7μm thick and 10μm long to 25μm wide, 22μm thick and 100μm long. These were made with the long axis in the [110] direction in the single crystal copper, Figure 1 c.

For the study into fracture properties the microcantilevers were machined so as to contain a single grain boundary approximately 1μm from the fixed end. Cantilevers had a pentagonal cross section as used in the fracture studies of Di Maio and Roberts [1] so that cracks would initially propagate with constant width. Cantilevers with a width of 4.5μm length, thickness of 5μm and length of 20μm were made with a Zeiss Nvision40. The undercutting was performed at an angle of 30°. A sharp notch was machined on the grain boundary in a single pass using a beam

8

current of 10pA for 60 seconds; this can be seen in figure 1a. This notch was found to be 700-800nm deep. Each cantilever was imaged using an SEM before testing to allow the width and thickness to be measured. EBSD was used to measure the misorientation across the grain boundary, and the boundary plane.

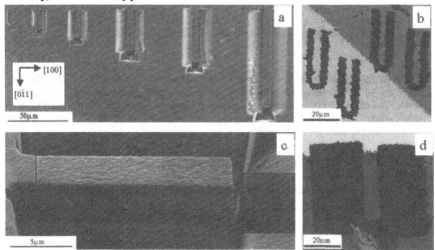

Figure 1: a) 6 cantilevers of differing size for measuring size effect in yield stress. b) EBSD showing single crystal cantilevers cut from polycrystalline copper for measuring anisotropy in Young's modulus. c) Microcantilever with pentagonal cross section and sharp notch at grain boundary for measuring fracture toughness. d) EBSD of cantilever for fracture toughness measurements showing position of grain boundary.

Testing Of Microcantilevers

All the microcantilevers were tested using a AFM/nanoindenter (MTS nano XP). Cantilevers were first located using the optical microscope and then scanned to produce a topographical map of the area containing the cantilever. A second higher resolution scan was produced of the area immediately around the microcantilever, allowing the indenter tip to be placed close to the free end of the microcantilever and the cantilever to be loaded. For measuring elastic properties, multiple loadings were performed at decreasing distances from the fixed end of each cantilever, at a constant strain rate of $6 \times 10^{-4} s^{-1}$ to displacements of 200nm. To measure size effects in yield stress each cantilever was loaded once at a rate of 5nms^{-1}; depending on the cantilever size the target displacement was set between 400nm and 3μm to ensure that yielding occurred. For measuring fracture toughness of grain boundaries a single displacement of each cantilever was carried out at a strain rate of $6 \times 10^{-4} s^{-1}$ until fracture occurred or the displacement reached 1000nm. Following testing each cantilever was imaged in the SEM to allow the effective cantilever length (i.e. the distance between the loading point and fixed end) to be measured.

DISCUSSION

Single Crystal Elastic Properties in Polycrystalline copper

The cantilevers tested were analyzed using the methodology outlined in [5]. In summary it was found that for beams with an aspect ratio (length from loading point to fixed end, L: width) greater than 6, simple encastré beam behavior was followed, so that:

$$S = \frac{\delta}{P} = \frac{L^3}{3EI} \tag{1}$$

Here S is the measured compliance (the ratio between displacement, δ, and load, P), E is the Young's modulus along the length of the beam and I is the beam's second moment of area. Young's modulus was calculated from the gradient of plots of L^3 versus S. The expected value of Young's modulus for a given orientation was also calculated from single crystal elastic constants in the literature [8] using the relationship:

$$\frac{1}{E_{uvw}} = \frac{c_{11} + c_{12}}{(c_{11} - c_{12})(c_{11} + 2c_{12})} - 2[\frac{1}{(c_{11} - c_{12})} - \frac{1}{2c_{44}}](u^2v^2 + v^2w^2 + w^2u^2) \tag{2}$$

Table 1 summarizes the values measured for the cantilevers of orientation [uvw] manufactured inside single grains with surface plane (hkl) and the literature-derived values of E.

(hkl)	[uvw]	Experimentally measured Modulus (GPa)	Modulus range derived from Literature values (GPa)
8 $\overline{13}$ 23	11 5 $\overline{1}$	87	85-103
8 $\overline{13}$ 23	11 5 $\overline{1}$	79	85-103
$\overline{7}$ 8 9	15 12 1	131	116-134

Table 1: Values of Young's modulus both experimentally measured using microcantilevers and derived from the literature values of elastic constants.

The experimentally measured values are in good agreement with those derived from literature values for single crystal Young's Modulus.

Yield Stress Size Effects in Single Crystal Copper

Each load-displacement curve was converted to a stress-strain curve using simple beam theory. The maximum stress (at the lower surface at the beam root) in a triangular cantilever loaded in the elastic regime is:

$$\sigma_{max} = \frac{24PL}{ah^2} \tag{3}$$

and the maximum strain is:

$$\varepsilon_{max} = \frac{2h\delta}{L^2} \tag{4}$$

Where P=load, L=effective beam length, a=beam width, h=beam thickness and δ=displacement.

Figure 2a shows three stress-strain curves for cantilevers of differing thickness. As the beam thickness is deceased the yield stress increases from 300MPa for the largest cantilever to 650MPa for the smallest. The yield stress for all cantilevers tested is plotted in figure 2b. Above a thickness of 5μm the yield stress is approximately 300MPa. As cantilever size is reduced below this value the yield stress rapidly increases towards almost 1GPa. These values are in good agreement with results from Kiener et al. [9] for both micro-compression and rectangular microcantilever tests. They observed an increase in yield stress from around 300MPa for their largest specimens, towards 950MPa for the smallest.

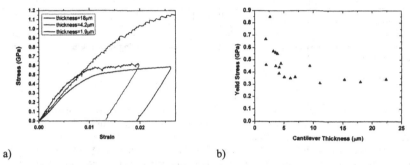

a)   b)

Figure 2. a) Load-displacement curves for three cantilevers of differing size. b) Yield stress as a function of beam thickness showing clear size effect as beam thickness is decreased.

Fracture Properties of Grain Boundaries

Four microcantilevers were tested in Cu-0.02wt%Bi: two on a Σ7 boundary (GB1) and two on a high angle grain boundary of misorientation 28° (GB2). Fracture was seen to occur in each case, the displacement rising rapidly and unstably at a critical applied load (Figure 3a). After testing each fracture surface was imaged using a FEG-SEM, Figure 3b and 3c.

a)   b)   c)

Figure 3; a) Load-displacement curves for testing of four-grain boundaries in copper bismuth. b) Tested cantilever showing fracture at the grain boundary. c) Fracture surface.

Fracture toughness was calculated using the procedure used by Di Maio and Roberts [1]. The results are summarized in Table 2. The vector given in "misorientation" is the common vector in the two grains about which the misorientation rotation occurs; "plane 1" and "plane 2" are the grain boundary planes in the two grains. The different grain boundaries have significantly different measured fracture toughness values, though their range is comparable to the 2-4 $MPam^{0.5}$ reported by Wang and Anderson [10] for boundaries in $\Sigma 9$ and $\Sigma 11$ bi-crystals.

Boundary	Misorientation	Plane1	Plane2	Fracture Toughness $MPam^{0.5}$
1	$42^\circ <7\ 8\ 12>(\Sigma 7)$	$\{1\ 2\ 1\}$	$\{1\ 30\ 1\}$	0.99
1	$42^\circ <7\ 8\ 12>(\Sigma 7)$	$\{1\ 2\ 1\}$	$\{1\ 30\ 1\}$	1.08
2	$28^\circ <7\ 19\ 13>$	$\{10\ 34\ 3\}$	$\{1\ 28\ 10\}$	1.50
2	$28^\circ <7\ 19\ 13>$	$\{10\ 34\ 3\}$	$\{1\ 28\ 10\}$	1.51

Table 2: Grain boundary misorientation and fracture toughness measured using microcantilever testing for two boundaries in Cu-0.02wt%Bi

CONCLUSIONS

Cantilevers machined into single grains of polycrystalline copper have been used to measure single crystal Young's modulus values. The values measured are in good agreement with those predicted from literature values in bulk materials.

Cantilevers ranging from $1.7\mu m$ thick by $12\mu m$ long to $22\mu m$ thick by $100\mu m$ long were machined with the long axis in the same crystallographic direction in single crystal copper. The yield stress increased rapidly with decreasing beam cross-section.

Single grain boundaries in an embrittled copper-bismuth alloy have been isolated in microcantilevers and a sharp notch placed on the boundary by FIB machining. By loading the boundaries to failure the fracture toughness of individual grain boundaries was determined. A larger range of boundaries are currently being studied and related to local chemistry characterized with TEM-EDX. The testing of the susceptibility of grain boundaries to stress corrosion cracking in steel is also being investigated [11].

These exploratory tests show the potential of using microcantilevers cut into the surface of single crystal polycrystalline materials for testing local mechanical properties.

REFERENCES

1. D. Di Maio and S.G. Roberts, *Measuring fracture toughness of coatings using focused-ion-beam-machined microbeams,* Journal of Materials Research, **20**(2): p. 299-302. 2005.
2. D. Kiener, W. Grosinger, G. Dehm, and R. Pippan, *A further step towards an understanding of size-dependent crystal plasticity: In situ tension experiments of miniaturized single-crystal copper samples,* Acta Materialia, **56**(3): p. 580-592. 2008.
3. M.D. Uchic and D.A. Dimiduk, *A methodology to investigate size scale effects in crystalline plasticity using uniaxial compression testing,* Materials Science and Engineering a-Structural Materials Microstructure and Processing, **400**: p. 268-278 .2005.
4. F.M. Halliday, D.E.J. Armstrong, J.D. Murphy, and S.G. Roberts, *Nanoindentation and micromechanical testing of iron-chromium alloys implanted with iron ions,* Adavnced Materials Research, **59**: p. 304-307. 2009.

5.   D.E.J. Armstrong, A.J. Wilkinson, and S.G. Roberts, *Measuring Anisotropy in Young's Modulus of Copper Using Microcantilever Testing,* Journal of Materials Research (submitted). 2009.

6.   W. Sigle, L.S. Chang, and W. Gust, *On the correlation between grain-boundary segregation, faceting and embrittlement in Bi-doped Cu,* Philosophical Magazine a-Physics of Condensed Matter Structure Defects and Mechanical Properties, **82**(8): p. 1595-1608. 2002.

7.   V.J. Keast, A. La Fontaine, and J. du Plessis, *Variability in the segregation of bismuth between grain boundaries in copper,* Acta Materialia, **55**(15): p. 5149-5155. 2007.

8.   H. Ledbetter and E.R. Naimon, *Elastic Properties of Metals and Alloys. II. Copper,* Journal of physical and chemical reference data, **3**: p. 897. 1974.

9.   D. Kiener, C. Motz, T. Schoberl, M. Jenko, and G. Dehm, *Determination of mechanical properties of copper at the micron scale,* Advanced Engineering Materials, **8**(11): p. 1119-1125. 2006.

10.   J.S. Wang and P.M. Anderson, *Fracture-Behavior of Embrittled Fcc Metal Bicrystals,* Acta Metallurgica Et Materialia, **39**(5): p. 779-792. 1991.

11.   D.E.J. Armstrong, M. Rogers, and S.G. Roberts, *Micromechanical Testing of Stress Corrosion Cracking of Individual Grain Boundaries,* Journal of Materials Research (submitted). 2009.

Mater. Res. Soc. Symp. Proc. Vol. 1185 © 2009 Materials Research Society 1185-II04-02

Deformation Mapping in Micro- and Nanoscale Fibers

Christina Garman[1], Natalie Bindert[2], Adhira Sunkara[3,4], Leocadia Paliulis[5], and Donna M. Ebenstein[3]

[1]Computer Science Department, Bucknell University, Lewisburg, PA 17837, U.S.A.
[2]Mechanical Engineering Department, Bucknell University, Lewisburg, PA 17837, U.S.A.
[3]Biomedical Engineering Department, Bucknell University, Lewisburg, PA 17837, U.S.A.
[4]Biomedical Engineering Department, Washington University, St. Louis, MO 63130, U.S.A.
[5]Biology Department, Bucknell University, Lewisburg, PA 17837, U.S.A.

ABSTRACT

Small-scale natural fibers are among the biological materials being studied by researchers seeking innovative methods to create new high performance materials. For example, spider dragline silk fibers are being studied because of their unique combination of high strength-to-weight ratio and high extensibility, which leads to a tough and lightweight fiber. Biomimetic fibers based on spider silk have been a focus of research for the past decade. However, there are still many unanswered questions about the mechanisms by which silk achieves its unique mechanical properties, as well as challenges in mechanical testing of biomimetic silk fibers (which is often hindered by both small diameters and limited material availability). A method to characterize local mechanical behavior in small diameter fibers was developed to both improve understanding of structure-property relationships in natural fibers and provide a method for comparing mechanical behavior in natural and biomimetic fibers. The deformation mapping technique described in this paper, which utilizes a piezoelectric micromanipulator with pulled glass tips, an inverted microscope with attached camera, and an image processing MATLAB program, is also applicable to the characterization of other micro- and nanoscale fibers where local deformation mechanisms may be of interest (e.g., for mechanical characterization of electrospun fibers).

INTRODUCTION

Small-scale natural fibers are among the biological materials being studied in the field of biomimetics, or applying lessons learned from nature to solve engineering problems. Natural fibers often have superior mechanical properties to traditional synthetic polymer fibers. For example, spider dragline silk fibers (with diameters of 5-10 microns) have a unique combination of high strength-to-weight ratio and high extensibility, which leads to a tough and lightweight fiber [1]. Such properties would be desirable to achieve for fibers to be used in clothing, sutures, and other applications requiring a strong, tough and lightweight fiber.

Biomimetic fibers based on spider silk have been a focus of research for the past decade [1]. However, there are still many unanswered questions about the mechanisms by which silk achieves its unique mechanical properties. Additionally, as researchers work to develop biomimetic fibers, mechanical testing is often hindered by both small diameters and limited availability of the experimental materials. For example, many biomimetic fibers are generated through the electrospinning process, which can result in sub-micron diameter fibers [2].

15

To address current and future issues in the field of biomimetic fibers, a method to characterize local mechanical behavior in small diameter fibers was developed. The goal of this research is to develop a tool that can be used both to improve understanding of structure-property relationships in natural fibers and to provide a method for comparing mechanical behavior in natural and biomimetic fibers. The deformation mapping technique described in this paper, which utilizes a piezoelectric micromanipulator equipped with a pulled glass tip, an inverted microscope with attached camera, and an image processing MATLAB program, is also applicable to the characterization of other micro- and nanoscale fibers where local deformation mechanisms may be of interest.

EXPERIMENTAL METHODS AND SAMPLE RESULTS

The deformation mapping technique described in this paper utilizes a piezoelectric micromanipulator equipped with a pulled glass tip to pull small diameter fibers, an inverted microscope with attached camera to capture images of the fibers during the pulling process, and an image processing MATLAB program [3] to analyze the resulting deformations and strains in the small diameter fibers. The following sections will provide detailed descriptions of the equipment used, tip preparation, sample preparation including an optional dyeing procedure to label regions of the fiber, the fiber stretching procedures, and the data processing methods.

Equipment

These experiments were performed using the Zeiss IM-35 inverted phase contrast microscope (Oberkochen, Germany) and the joystick-controlled custom-built piezoelectric Ellis micromanipulator described in detail by Zhang and Nicklas [4]. Images from the attached Panasonic WV-BD400 video camera were acquired through a Flashbus card in a Dell Dimension 4550 computer. However, similar experiments could be performed using any inverted optical microscope combined with a micromanipulator with sufficiently precise and smooth motion.

Tip preparation

The micromanipulator was equipped with a micropipette holder that can accommodate cylindrical probes or needles. Experiments were performed using 1 mm outer diameter glass needles pulled to sharp tips (with nominal diameters of 0.1 μm) following the protocol of Zhang and Nicklas [4]. However, the sharp tip is needed only for the optional dying of the fiber. The pulling of the fiber could be performed with pre-drawn glass needles or probes made from other stiff materials.

To facilitate using the tip to apply dye to a fiber, the glass tips were treated to make them hydrophilic or hydrophobic, depending on the dye to be used. Hydrophilic tips were used for water-based dyes, while hydrophobic tips were used for oil-based dyes. To create hydrophilic tips, glass needles were placed in a plasma oxygen chamber for 10 minutes to populate the surface with hydroxyl groups. For hydrophilic tips, after the plasma oxygen treatment the glass needles were soaked in a 1.0 mM solution of 3-(trichlorosilyl) propyl methacrylate (TPM) in dry ethanol to create an hydrophobic TPM monolayer on the surface.

Figure 1: Silk fiber sample preparation. (a) Silk fiber glued to cover slip and attached to glass slide with 2 cm diameter hole (bar = 1 cm). (b) Example of "band-stained" *B. mori* silkworm silk fiber, also showing a pool of Safranin O dye in the top of the image (bar = 50 μm). (c) Example of "dot-stained" *N. arabesca* spider silk fiber, stained with Safranin O (bar = 8 μm).

Sample preparation

Spider silk fibers between 0.5 and 1 cm long were harvested from the web of a *Neoscona arabesca* spider. Silkworm silk fibers of the same length were harvested from *Bombyx mori* cocoons using forceps. To prepare silk for stretching, a viscous cyanoacrylate-based adhesive (Loctite 454) was used to glue the harvested fiber to a 2.5 cm x 2.5 cm cover slip at two points, constraining the fiber with little or no slack. After the adhesive had dried, the cover slip was attached to a glass slide with a 2 cm diameter hole using vacuum grease (Dow Corning, Midland, MI). This method allows for a hydrating fluid to be placed in the hollow in the slide, if sample hydration is desired, and reduces the cost of the disposables, since the slide can be reused. However, if no hydration is necessary, fibers can be glued directly to a glass slide or other transparent platform. A mounted silk fiber is shown in Figure 1a. After the fiber was secured to the slide, the slide was mounted on the microscope stage and an objective lens was selected that allowed observation of a large region of the fiber but with sufficient resolution to see some internal detail of the fiber. For spider silk, we found that a 100X 1.25 N.A. oil immersion objective provided suitable resolution, while for thicker silkworm silk a 16X objective was used.

If there was not sufficient texture within the fiber to track deformations, or if more regularly spaced points were desired, the micromanipulator equipped with a sharp glass needle was used to apply dye to the fiber. To do this, a small amount of dye was placed directly on the cover slip near the fiber, as shown in Figure 1b. The tip of the micromanipulator-controlled treated glass needle was dipped into the dye, and then the dye covered tip was gently brought into contact with the fiber to transfer the dye. The fiber can either be "band-stained" to coat regions of the fiber (as shown in Figure 1b) or "dot-stained" to attach discrete dye particles to the fiber (as shown in Figure 1c). For the fibers shown in Figure 1, a hydrophobic TPM-coated tip was used to stain the fibers with Safranin O, a histology dye used to stain lipids.

Fiber stretching protocol

To stretch the mounted fiber, the micromanipulator was used to position the tip of the glass needle next to the fiber. The probe was then moved laterally to stretch the fiber in a V-Pattern, as shown in Figure 2a, while capturing images using the camera attached to the microscope. The stretching motion was slow enough that images could be captured with only small amounts of displacement between sequential images to facilitate automated image analysis.

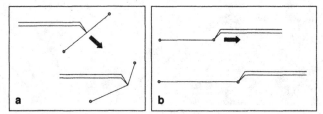

Figure 2: Schematic illustrating methods of deforming the fiber. Arrows indicate the direction that the needle is moved to deform the fiber. (a) Illustration of V-Pattern deformation, with both ends of the fiber glued to the cover slip. (b) Illustration of uniaxial deformation, with one end of the fiber glued to the cover slip and the other end of the fiber glued to the glass needle.

A sequence of images was collected during V-Pattern stretching of a band-stained *B. mori* silkworm silk fiber. Representative images from the sequence (Figure 3) show that the stained region of the fiber is moving down and to the left relative to the fixed dye pool. An alternate stretching technique currently under development to facilitate image analysis is shown in Figure 2b. Using this method, only one end of the fiber is glued to the cover slip; the other end of the fiber is glued to the glass needle. This method allows for uniaxial stretching of the sample. While this method facilitates data analysis, it requires more handling of the fiber samples.

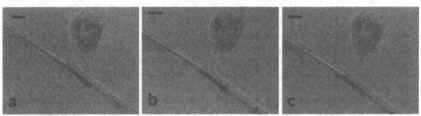

Figure 3: Sample images collected while stretching a band-stained *B. mori* silkworm silk fiber using the V-Pattern method. Bar = 50 μm. Pixel size = 714 nm.

Data analysis

The sequence of images collected during V-Pattern stretching of a band-stained silkworm silk fiber (Figure 3) was analyzed using a MATLAB program developed by the Hemker group at The Johns Hopkins University [3]. This program performs image analysis on a sequence of .TIF images to track features in the image and calculate deformations and strains based on user-defined points [3]. Data analysis using two different types of point arrays, a rectangular grid and linear sequence, is shown below in Figures 4 and 5.

Figure 4a shows a rectangular grid with 30 pixel spacing between each pair of points overlaid onto the image from Figure 3a. By analyzing pixel displacement using this rectangular grid, which includes points inside and outside the fiber, the motion of the fiber can be quantified. The MATLAB program tracks the movement of a group of pixels surrounding each point in the grid. In the 3D plots in Figures 4b and 4c, the x and y axes correspond to the x-y coordinates of

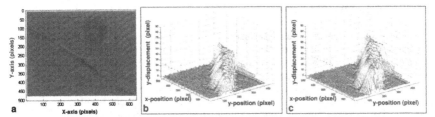

Figure 4: Quantification of pixel motion using a grid. **(a)** Rectangular grid used for image correlation analysis. **(b)** Y-axis pixel displacements for Figure 3b. **(c)** Y-axis pixel displacements for Figure 3c. Displacements were computed relative to Figure 3a. Pixel size = 714 nm.

the grid points in Figure 4a, while the z-axis measures the vertical (y-axis) displacement of each point as the fiber was pulled. Figures 4b and 4c show that motion was detected only for the grid points within the fiber, and the displacement of these points increased though the sequence of images as the fiber was pulled. Similar analysis of pixel displacement in the x-direction showed that the pixels within the fiber were moving to the left, with peak displacements increasing through the sequence (data not shown). These results are consistent with the visible movement of the stained band in the image sequence in Figure 3.

To investigate strains within the fiber during stretching, analysis was performed using a linear sequence of points evenly spaced along the axis of the fiber. Figure 5 shows the motion of a line of 5 points spaced approximately 50 pixels apart. Figures 5b and 5c illustrate tracking of the 5 pixel regions through the sequence of images. The x- and y- coordinates of the tracked pixels were used to compute initial gage lengths ($L1_0$, $L2_0$, $L3_0$, and $L4_0$) and subsequent changes in segment length, ΔL. Strains were then were calculated as $\Delta L/L_0$ for each pair of points. Strain analyses performed over the whole fiber length and at different regions within the fiber showed no clear trends with increased stretching or location in the fiber. Based on these results, we conclude that the images in this sequence captured translation and possibly rotation of the fiber rather than straining, suggesting that the fiber was being reoriented during this initial stage of stretching. This could result, for example, from some initial slack in the fiber. Strain in the fiber likely occurred after the fiber moved out of the field of view of the camera.

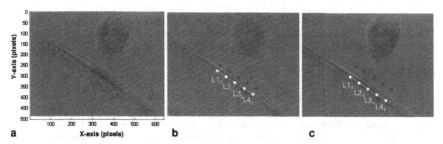

Figure 5: Strain mapping within the fiber using a linear grid. (a) Initial linear grid along the axis of the fiber. (b) Tracking of pixel movement for Figure 3b. (c) Tracking of pixel movement for Figure 3c. Distances used for strain analysis are labeled in each image. Pixel size = 714 nm.

19

DISCUSSION AND CONCLUSIONS

This study demonstrates a mechanical testing technique requiring minimal sample preparation that can be used to measure local deformations and strains within small-scale fibers. While this method is easy to implement and can provide valuable information about fiber motion, there are several limitations to the current analysis. First, although this study demonstrated the ability to track pixel motion, the images in this sequence failed to capture the part of the experiment during which strain was induced in the fiber. Second, with these small diameter fibers there is a trade-off between the observed fiber length and width. For these experiments, a longer length of fiber was selected to keep the fiber within the field of view longer, which limited the ability to analyze transverse strains within the fiber.

Enhancements are currently under development to address these and related issues. For example, the MATLAB program can perform automated strain mapping analyses, but works most effectively if fibers are being stretched horizontally and uniaxially. Switching from V-pattern stretching to uniaxial stretching (as described above) will address this problem and will facilitate capturing data throughout the stretching process without the fiber leaving the field of view of the camera during the process. Coupling this approach with a higher magnification will facilitate 2D mapping of local strains within the fiber. In addition to these changes in experimental methods, the system is being refined to facilitate the measurement of stress-strain curves and the extraction of material properties. First, we are replacing the joystick-controlled piezoelectric micromanipulator with a programmable, stepper-motor based micromanipulator (Sutter MP-285, Novato, CA) to achieve repeatable and quantifiable motion. In addition, we will add a force sensor to the system to monitor the force generated by the fiber during stretching.

In conclusion, this paper presents an easy-to-implement mechanical testing technique that allows mapping of local deformations in small diameter fibers with minimal sample preparation. Adding the programmable manipulator and a force sensor will lead to a powerful technique for displacement controlled mechanical testing of small diameter fibers. While demonstrated here for a silkworm silk fiber, this technique can also be applied to any small-scale natural, biomimetic, or synthetic fiber, including larger diameter electrospun fibers.

ACKNOWLEDGMENTS

The authors would like to thank Erin Jablonski's Nanofabrication Laboratory at Bucknell University for assistance with coating the glass needles, and the National Science Foundation (award EEC-0741487) for funding of the Nanofabrication Laboratory. C.G. and N.B. were funded by the Presidential Fellowship Program at Bucknell University.

REFERENCES

1. L. Romer and T. Scheibel. Prion 2(4): 154-161 (2008).
2. A. Greiner and J.H. Wendorff, Angew. Chem. Int. Edit. 46(30): 5670-5703 (2007).
3. C. Eberl, R. Thompson, and D. Gianola. *Digital Image Correlation and Tracking,* MATLAB Central, <http://www.mathworks.com/matlabcentral/fileexchange/12413> accessed Jan. 27, 2009.
4. D. Zhang and R.B. Nicklas in *Methods in Cell Biology, Vol. 61,* edited by C.L. Rieder (Academic Press, San Diego, CA, 1999), p. 209-218.

Mater. Res. Soc. Symp. Proc. Vol. 1185 © 2009 Materials Research Society 1185-II04-06

Methods to Measure Mechanical Properties of NEMS and MEMS: Challenges and Pitfalls

Ingrid De Wolf[1,2], Stanislaw Kalicinski[1,2], Jeroen De Coster[1], Herman Oprins[1]
[1]IMEC, Kapeldreef 75, Leuven, B-3001 Belgium
[2]K.U.Leuven, Dept. MTM, Leuven, B-3001 Belgium

ABSTRACT

This paper focuses on the measurement of material properties of micro and nano-electromechanical systems. Two different methods are discussed: electrical or optical measurements of the resonance frequency, and measurements of the Raman frequency shift. The main focus of this paper is on challenges and pitfalls related to the use of these techniques for the study of MEMS and NEMS.

INTRODUCTION

The main difference between micro- and nano electro-mechanical systems (MEMS, NEMS) and ICs is that the former can (or have to) move. Because of this motion, mechanical issues such as creep, fatigue, fracture, stress, stress gradient, yield strength etc. play a major role in determining their functionality and reliability. Also the ambient in which this motion takes place (normal air, vacuum, N_2 etc.), the temperature, the cleanness of this ambient (particles, outgassing, adsorption) plays a major role and can have a direct effect on the mechanical properties of these devices.

Measuring these properties is not an easy task. First of all, the assumption that the mechanical properties of a film are the same for the NEMS/MEMS made-out of this film is not necessary correct. Secondly, to test for example for fatigue, one should be able to monitor these systems during a very long time, while stress-cycling them. This is in general relatively easy if a fatigue sensitive parameter (for example the resonance frequency of a Si -resonator) can be measured through electrical means. It becomes more complicated if it has to be measured through optical means. Thirdly, drifts in resonance frequency, a parameter that is often used to check whether there are mechanical reliability issues, can also be caused by other mechanisms, such as charging, outgassing, adsorption, fly catching effect etc. Especially on the nano scale, this becomes important. How can it be ensured that the envisioned mechanical property is being monitored? And last, the risk that the measurement system interferes with the mechanical property to be measured is increasing with decreasing dimensions of the devices.

In this paper we will address these issues and propose possible solutions to the problems discussed.

RESONANCE FREQUENCY

Electrical measurements

The resonance frequency of NEMS or MEMS is often used to monitor possible changes in mechanical properties. Changes in resonance frequency were for example used in [1] to study fatigue in poly-crystalline Si MEMS using dedicated test structures. The vibration frequency can also change due to mass increase or decrease, for example caused by outgassing or absorption processes [2]. In addition, this can also be due to the so called 'fly catching effect' [3, 4]: Electrostatic charging of NEMS or MEMS can occur because of high frequency vibrations in air; this is probably a friction effect between the resonating beam and airborne particles, resulting in charging. As a result, airborne particles from the ambient air are themselves attracted by the vibrating device. These particles increase the mass of the resonator, resulting in a decrease in the resonance frequency [5]. Kazinczi demonstrated that only beams vibrating at resonance attract particles, while non-resonant neighbouring beams stay particle free [2, 3, 6]. This effect is also related to the amplitude of the vibration.

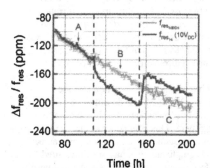

Time [h]

Figure 1. Relative change of the resonance frequency of the unsealed MEM resonator in vacuum at 300K determined at $V_{DC} = 10V$ (grey/yellow) and of the extracted unbiased resonance frequency f_{reso} (black). During period A no stress bias is applied. During period B 40V stress and during period C -40V stress is applied.

It is clear that the resonance frequency of MEMS/NEMS can be affected by changes in mechanical parameters: fatigue, probably also creep, mass changes due to oxidation, outgassing, particles trapping etc. However, the resonance frequency can also be directly affected by charging of a dielectric near or on the MEMS/NEMS. Such trapped charge can induce a build-in voltage, affecting the resonance frequency. In many electro-statically actuated NEMS/MEMS, such a dielectric is present. Kalicinski et al. showed that such charging effects can highly affect the resonance frequency [7]. An example is given in Fig. 1, where the relative change of the resonance frequency of the first in-plane vibration mode of an electrostatically actuated 1.5 μm thick, 6 μm wide crystalline Si beam resonator (made on SOI) is studied. The drift in resonance frequency (measured at 10V) is monitored in a vacuum chamber during more than 200 hours, without a DC stressing voltage applied to the device (during region A as indicated in Fig. 1), with a positive DC bias voltage (40V) applied to the device (region B, Fig. 1), or with a negative DC bias voltage (-40V, region C in Fig. 1). This stress voltage is much smaller than the pull-in voltage of this device (~ 300V). The resonance frequency drift measured during this experiment is shown by the black curve (f_{res14}) in this figure. The curve shows two effects: The resonance frequency decreases over time, and, in addition, it shifts down when applying 40V bias, and shifts up when applying -40V bias. There are clearly several effects taking place here. How can one distinguish between mechanically induced effects and electrically induced effects on the resonance frequency?

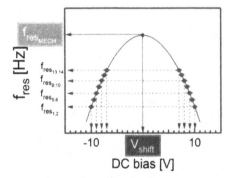

Figure 2. Measured resonance frequencies f_{res} versus DC bias. Line: Fit of model [7] to the points giving the mechanical resonant frequency f_{reso} and V_{shift}.

To distinguish between the two, Kalicinski proposed a measurement technique in which the resonance frequency is measured at different, both positive and negative voltages, as shown in Fig. 2. The obtained data are then fitted using an appropriate function [7]. Form this fit the maximum of the curve and the voltage at which this maximum is reached are deduced. The maximum of the curve, $f_{resMECH}$, corresponds then with the mechanical resonance frequency. Any change in this value is caused by mechanical changes of the device (including mass changes). The voltage, V_{shift}, indicates whether there is charging occurring that affects the resonance frequency. Using this measurement principle, the measured curve from Fig. 1 can be split into a curve showing the purely electrical effects on the resonance frequency, as given in Fig. 3, and a curve showing purely mechanical effects, as given in Fig. 1 ($f_{resMECH}$). This experiment shows that after correcting for the electrical changes, the resonance frequency still shows a steady down shift with time during this test. This was caused by outgassing of components present in the vacuum chamber, affecting the mass of the resonator. It is clear that the resonance frequency is a very good and sensitive parameter to study mechanical properties and changes in these properties. But this experiment demonstrated that also charging can affect the resonance frequency of a NEMS or MEMS device. Care has to be taken to distinguish between charging induced and mechanical changes of this resonance frequency.

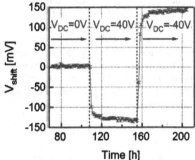

Figure 3. Extracted V_{shift} changes during testing with DC stressing. The date are extracted from Fig. 1 using information from Fig. 2.

Optical measurements

In the above discussed experiment, the device was actuated electrically and the resonance frequency was measured using electrical means. It is also possible to actuate mechanically, by using a shaker, and measure electrically; or to actuate electrically or mechanically and measure optically. The instrument that is best suited for the latter is a Laser Doppler Vibrometry (LDV). Figure 4 shows an experiment performed by De Coster. In this experiment, the resonance frequency of the torsion and bending mode of a 8μmx8μm SiGe micromirror [8] was first calculated using simulations (Fig. 4, left), and then measured using an LDV system mounted on a vacuum probe station (Fig. 4, right). The combination of FEM and experiment allowed to fine-tune the FEM results and to deduce the E-modulus of the mirrors. This principle can be applied to any MEMS/NEMS of which the resonance frequency can be deduced.

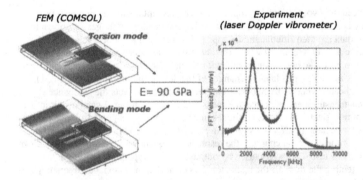

Figure 4. Left: Finite element model showing the torsion and bending mode of a SiGe micro-mirror. Right: LDV measurement of the resonance frequencies of these modes. The combination of FEM and experiment allowed deducing the E-modulus of the mirrors.

The LDV can also be used to study the mechanical reliability of NEMS and MEMS by monitoring optically the resonance frequency over time. An example of the results of such an experiment, performed on the same micromirrors as discussed in Fig. 4, is shown in Fig. 5b. In this experiment, the mirrors were cycled (tilted) up to nearly 3×10^{12} times. The resonance frequency was deduced at regular intervals from the step-response of the mirrors, as shown in Fig. 5a. The aim of the experiment was to study whether any fatigue was present in these SiGe mirrors. In this experiment, no drift in resonance frequency was measured, indicating no fatigue problems [8,9].

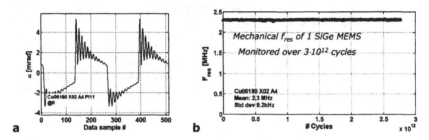

Figure 5. Left: Step response of a SiGe micromirror measured in vacuum using LDV. Right: Resonance frequency obtained from the left figure, and monitored over time.

A typical LDV set-up has a vertical resolution which is in the pm range, and can for this reason in many cases also be applied to NEMS, as long as part of the focused laser light reflects back to the detector. However, an important issue making such measurements difficult with common set-ups, is drift of the measurement set-up. The typical drift in time of probe stations is several microns in X-Y direction, and about 1 micron in Z (out-of-focus). This should be corrected for if one wants to monitor NEMS (or MEMS) during a long time using optical means. For the experiment shown in Fig. 5b, we added an autofocus systems and image correlation based feed-back loops to the system to keep the probing laser in focus, and at the same position on the same pre-defined spot(s) on device. The continuous feed-back corrections in X,Y and Z keep the system within 1 μm on the same spot, which should be good enough to allow applying the same system to NEMS [9].

Another problem that might play a role for the study of NEMS with optical systems is the resonance frequency: The smaller the devices, the higher the resonance frequency. Our current LDV system is limited to frequencies of 20 MHz. However, systems based on heterodyne interferometers are being developed that can measure up to 1.2 GHz [10, 11]. Such systems have been used before to measure mechanical vibrations up to the GHz regime with sub-nm resolution [12], but they have poor amplitude accuracy because it is impossible to quantify all influences to the uncertainty budget. On the other hand, the heterodyne interferometer presented in [10] can be calibrated and the uncertainty quantified by the ratio of carrier frequency to the side bands generated by the phase modulation from a vibrating specimen. This was successfully applied to a MEMS resonator in [11], showing that these systems are promising for the study of NEMS.

RAMAN SPECTROSCOPY

The risk that the measurement system interferes with the mechanical property to be measured is increasing with decreasing dimensions of the devices. Micro-Raman spectroscopy, for example, is a very interesting technique to measure local mechanical stress or temperature in Si devices [13,14]. It can easily be applied to structures that are smaller than the probing beam spot. For example, already in 1997 [15] we showed that the phase of 250 nm wide, 45 nm thick TiSi lines could be studied with this technique. Several papers reported on stress measurements in MEMS using micro-Raman spectroscopy [16,17], and it is even used to study motion induced local strain (dynamic strain) in high frequency NEMS/MEMS [18, 19]. It is clearly a very interesting technique to study mechanical properties of NEMS or MEMS.

There is however one large drawback when using this technique on nanostructures: the probing laser beam of the Raman system will always induce local heat in the devices. The relation between Raman frequency shift ($\Delta\omega$), i.e. the difference between measured (ω) and stress-free frequency (ω_o), and temperature (T) [14]; and between Raman frequency shift and stress (σ) (assuming uniaxial or biaxial stress) [13] for crystalline Si is given respectively by:

$$\Delta\omega \ (cm^{-1}) = -0.0242 \ \Delta T \ (K) \qquad \text{where } \Delta\omega = \omega - \omega_o \qquad (1)$$

$$\Delta\omega \ (cm^{-1}) = -0.0023 \ \sigma \ (MPa) \text{ or } \Delta\omega \ (cm^{-1}) = -0.0023 \ (\sigma_{xx} + \sigma_{yy}) \ (MPa) \qquad (2)$$

Even if there might be some discussions on the accuracy of the last digits of the numbers in these relations, it is clear that the effect of temperature on the Raman frequency shift is much larger than the one of stress. An increase of the local temperature with only 1 °C induces a similar shift in the Raman frequency as what 10 MPa tensile stress would do. Fig. 6 shows a simple experiment in which the Si Raman frequency shift was measured as a function of the focused laser beam power incident on a (100) Si wafer. This shows that a power of 3 mW already gives a temperature induced down shift of the Raman peak corresponding with a tensile stress of 10 MPa. And such power levels are easily reached in typical micro-Raman spectroscopy experiments. This experiment was done on bulk silicon. In the case of a MEMS, the effect is much larger because there is less Si available to act as heat sinc.

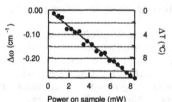

Power on sample (mW)

Figure 6. Shift of the Raman frequency of crystalline silicon with increasing power of the probing laser (458 nm) and the corresponding local temperature increase in silicon (Eq. (1)).

One can argue that this might be neglected if the laser-induced heating of the sample is the same all over the sample, so, any variations of the Raman frequency measured across the sample would indicate stress variations. However, this is not the case for NEMS/MEMS. These devices are mostly somewhere anchored to the substrate, and these anchor points act as heat sinks. An example of this was presented by van Spengen et al. [20] for 20 μm wide poly-Si cantilever beams (Fig. 7). The intention of this experiment was to measure the stress in these beams using micro-Raman spectroscopy. A large downshift of the Raman frequency was found, corresponding (if due to stress) with tensile stress values up to 400 MPa (assuming uniaxial stress, Eq. 1). This maximal stress value is the same for the three longest beams. It is clear that the measured Raman shift in this experiment is not caused by stress, but by local heating of the beams by the focused laser light. Close to the bond pads at the left and right side of the beams, the heat can get away and the measured Raman frequency shift is small. Closer to the center of the beam, the heat cannot get away easily and the heating effect is maximal. In fact, one could use this effect as a kind of failure analysis tool to check whether beams are stuck or not stuck to the substrate. In the former case they would show less heating-induced Raman frequency shift.

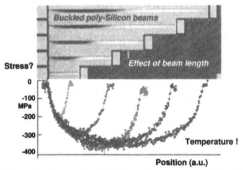

Figure 7. Stress calculated from the measured shift of the Raman frequency of poly-crystalline silicon beams with different lengths between 150 μm (shortest beam) and 750 μm (longest beam) ('Position' scale in arbitrary units, a.u.). The stress is calculated using Eq. (2) assuming (incorrectly) that the measured shift is due to stress. The shift is caused by local temperature increase in the beams caused by the probing laser (458 nm).

The thickness of the beams used in that experiment is not known. So, a question that rises is: For which sample thicknesses local heating starts influencing the Raman shift? This of course not only depends on the thickness, but also on the width of the investigated structures and on the materials properties. From a Raman experiment on a thinned crystalline Si sample (it was thinned for X-section TEM analysis), we deduced that beam induced heating increases fast if the membranes becomes thinner than 3 μm [21].

Figure 8 shows the variation of the Raman peak position (Raman frequency, ω) with distance from the hole in the TEM sample. The measurement was done with about 1 mW power on the sample. On top of the figure, a shematic drawing indicates the position of SiO_2 trenches which were present in the sample. The straight line (y-axis) in the figure indicates the thickness variation of the sample, deduced from theoretical considerations and experimental values of the Raman intensity variations. We clearly see two kinds of variation of ω: A local variation near the trenches, due to the local stress variation near these structures; and an overall decrease of ω towards the hole in the sample starting at about 100 μm away from the hole. This is not due to stress variations, but to changes in the local temperature of the sample due to the probing beam. The position where this decrease starts corresponds with a sample thinckness of about 3 μm.

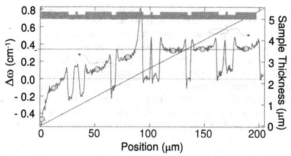

Figure 8. Left: Raman frequency shift measured on the X-section of a sample prepared for TEM inspection. The measurement is performed near the surface of a silicon sample with local SiO_2 trenches (as indicated in the schematic drawing at the top). The sample thickness (right axis, straight line) increases with position.

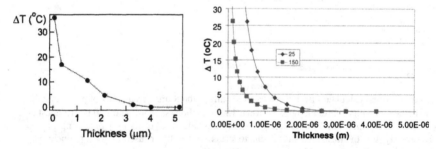

Figure 9. Left: Probing laser (from Raman spectrocope) induced temperature change in crystaline silicon in function of the sample thickness. The data are obtained from the experiment shown in Fig. 8. The laser power was ~ 1 mW and focus diameter ~1 μm. Right: Analytical calculation of the temperature induced by a local heat spot (1μmx1μm square) of 1 mW on a crystalline Si (thermal conductivity 150 W/m-k)) and a polycrystalline Si (thermal conductivity assumed 25 W/m-K) membrane in function of membrane thickness.

So, the decrease of the Raman frequency close to the hole in the TEM sample is predominantly due to local heating of the thin sample, i.e. due to a temperature increase. In Figure 8 seven points (bullets) in trench free areas are indicated. We assume that the stress in these points is the same (same distance from trenches). So, the difference in Δω between these points is due to a change in local heating because of sample thinning. Using equation (1), this heating can be calculated as a function of the sample thickness. The results is shown in Fig. 9a. This shows that for thicknesses smaller than 3 um, the temperature induced by the laser beam in the Si sample increases very fast and its effect will dominate any stress induced changes. One can expect that this effect becomes worse in materials with a larger absorption coefficient or more defective materials, such as polycrystalline Si, which is often used in NEMS. In addition, these calculations were done for membranes: For beams, the effect will be even larger. It is of course also possible to calculate these, using analytical relations as given for example in [22]. We did these calculations assuming a (square) beam spot of 1 μm, with 1 mW power, incident on a square crystalline Si (thermal conductivity assumed 150 W/m-k)) and polycrystalline Si (thermal conductivity assumed 25 W/m-K) membrane with different thicknesses. The results are shown in Fig. 9b. The trend and even the values indicated by these simulations correspond well to our roughly determined experimental results. In conclusion, one should be very carefull with the interpretation of the data when using micro-Raman spectroscopy to extract values of stress in MEMS. This becomes even more of a concern for NEMS. Local heating caused by the probing laser beam can falsify the measurement results.

CONCLUSIONS

This paper focused on two different methods that can be used to deduce information on mechanical material properties of MEMS and NEMS: Resonance frequency measurements and Raman spectroscopy. These methods are relatively simple and do not require SPM-based probing techniques or direct contact with the samples. However, they have some drawbacks.

The first method consists in the measurement and eventually monitoring of the resonance frequency of MEMS or NEMS. This can be done through electrical measurements, if the sample can be electrically activated and the resonance frequency electrically monitored. One drawback of this experiment is that the resonance frequency not only depends on mechanical parameters, it also depends on charges that can be trapped in or near the device. A methodology is presented that allows splitting mechanical and electrical effects in order to obtain the true mechanical resonance frequency.

It is not always possible to measure the resonance frequency of devices through electrical means. In that case, a laser Doppler vibrometer (LDV) can be used. However, this requires an autofocus system and X-Y drift compensation in such a way that the laser remains focused on the DUT. This is especially required when longer duration reliability tests have to be done. The LDV systems are limited in frequency, which might be an issue when testing NEMS. However, new systems are under development that should allow testing up to the GHz range.

The second method is the well known micro-Raman spectrometry method. The Raman frequency of silicon offers information on local mechanical strain in the sample. The technique is often applied to MEMS and even NEMS, but there is one important pitfall: the probing laser of the Raman system induces local heath in the sample, which in turn results in a shift of the Raman frequency. For silicon this effect can falsify stress-measurements for samples with a thickness smaller than 3 μm. In that case, differentiating between temperature effects and stress/strain is not that easy. And the effect is increasing for decreasing sample sizes, i.e. it is certainly an issue for NEMS. One can reduce the effect by measuring with a scanning laser and reducing the laser power as much as possible, but it can not be neglected and care has to be taken in interpretation of the Raman spectroscopy data.

REFERENCES

1. J.A. Connally and S.B. Brown, "Slow crack growth in single-crystal silicon," *Science* **256** (5063), 1537-1539 (1992).
2. R. Kazinczi, "Reliability of micromechanical thin-film resonators," PhD thesis, 'Technische Universiteit Delft', (Elburon Publishers) (2002).
3. R. Kazinczi, J.R. Mollinger and A. Bossche, "Environmental induced failure modes of thin film resonators," *Proc. SPIE, Smart materials and MEMS*, **4234**, 258-268 (2000).
4. R. Kazinczi, J.R. Mollinger and A. Bossche, "Adsorption-Induced Failure Modes of Thin-Film Resonators," *Proc. MSR Fall Meeting*, symposium L, L8.8 (2001).
5. N. Umeda, S. Ishizaki, and H. Uwai, "Scanning attractive force microscope using phoyothermal vibration," *J. Vac. Sci. Technol.* **B9**, 1318-1322 (1991).
6. J.A. Henry, Y. Wang and M.A. Hines, "Controlling energy dissipation and stability of micromechanical silicon resonators with self-assembled monolayers", *Applied Physics Letters* **84(10)**, 1765-1767 (2004).
7. S. Kalicinski, H.A.C. Tilmans, I. de Wolf, "A new impedance measurement based method to determine the mechanical resonance frequency and charging in electrostatically actuated MEMS," In: *18th Workshop on Micromachining, Micromechanics and Microsystems*, 16-18 September 2007; Guimaraes, Portugal, 329-332 (2007
8. L. Haspeslagh, J. De Coster, O. Varela Pedreira, I. De Wolf, B. DuBois, A. Verbist, R. Van Hoof, M. Willegems, S. Locorotondo, G. Bryce, J. Vaes, B. van Drieenhuizen and A. Witvrouw, "Highly reliable CMOS-integrated 11MPixel SiGe-based micro-mirror arrays for high-end industrial applications," In proc.: *Technical Digest International Electron Devices Meeting - IEDM*, 655-658 (2008)
9. I. De Wolf, J. De Coster, O. Valera Pedreira, L. Haspeslagh, A. Witvrouw, "Wafer level characterization and failure analysis of microsensors," In: *7th IEEE Conference on Sensors*, 144-147 (2008)
10. C. Rembe, S. Boedecker, A. Draebenstedt, F. Pudewills and G. Siegmund, „Heterodyne Laser-Doppler Vibrometer with a Slow-Shear-Mode Bragg Cell for Vibration Measurements up to 1.2 GHz," In proc. *SPIE* **7098** , 70980A-70980A-12 (2008).
11. S. Stoffels, S. Boedecker, R. Puers, I. De Wolf , H.A.C. Tilmans and C. Rembe, "Measuring the mechanical resonance frequency and quality factor of MEMS resonators with well-defined uncertainties using optical interferometric techniques," Accepted for *Transducers* (2009).
12. J. V. Knuuttila, P.T. Tikka, and M.M. Salomaa, "Scanning Michelson interferometer for imaging surface acoustic wave fields", *Optics Letters*, **25**, 613-615 (2000).
13. I. De Wolf, "Micro-Raman spectroscopy to study local mechanical stress in silicon integrated circuits," *Semicond. Sci. Technol.*, **11**, 139-154 (1996).
14. I. De Wolf, "Semiconductors," *Analytical applications of Raman spectroscopy*, ed. M. Pelletier (Blackwell Publishing, 1999), Chapter 10
15. I. De Wolf, D.J. Howard, M. Rasras, A. Lauwers, K. Maex, G. Groeseneken and H.E. Maes, "A reliability study of titanium silicide Lines using micro-Raman spectroscopy and emission microscopy," *Microelectron. Reliab.*, **37** **(10/11)**, 1591-1594 (1997).

16. VT Srikar, AK Swan, MS Ünlü, BB Goldberg and SM Spearing, "Micro-Raman measurement of bending stresses in micromachined silicon flexures,: *IEEE J. Microelectromech. Systems*, **12** (**6**), 779-787 (2003).

17. LA Starman Jr, EM Ochoa, JA Lott, MS Amer, WD Cowan and JD Bushbee, "Residual stress characterization in MEMS microbridges using micro-Raman spectroscopy," *Modeling and Simulation of Microsystems*, ISBN 0-9708275-7-1, 314-317 (2002).

18. John Hedley, Zhongxu Hu, Isabel Arce-Garcia, and Barry J. Gallacher, "Mode Shape and Failure Analysis of High Frequency MEMS/NEMS using Raman Spectroscopy," *Proceedings of the 3rd IEEE Int. Conf. on Nano/Micro Engineered and Molecular Systems* January 6-9, Sanya, China (2008).

19. EPSRC grant EP/C015045/1 (DyMARS) http://gow.epsrc.ac.uk/ViewGrant.aspx?GrantRef=EP/C015045/1

20. M van Spengen. Reliability of MEMS. PhD dissertation, ISBN 90-5682-504-6, Katholieke Universiteit Leuven (2004).

21. I. De Wolf, "Spectroscopic techniques for MEMS inspection", *Optical Inspection of Microsystems*, ed. W. Osten (Taylor&Francis, 2006), pp. 459-481

22. S. Lee, S. Song, V. Au, and K.P. Moran, "Constriction/Spreading Resistance Model for Electronic Packaging," *Proceedings of the 4th ASME/JSME Thermal Engineering Joint Conference*, **4**, 199-206 (1995).

Mater. Res. Soc. Symp. Proc. Vol. 1185 © 2009 Materials Research Society   1185-II04-08

UV Raman Spectroscopy Study of Strain Induced by Buried Silicon Nitride Layer in the BOX of Silicon on Insulator Substrates

V. Paillard, J. Groenen, P. Puech
CNRS-CEMES and Univ. Toulouse, 29 rue Jeanne Marvig, 31055 Toulouse Cedex 4, France

Y. Lamrani, M. Kostrzewa, J. Widiez, J.-Ch. Barbé, Ch. Deguet, L. Clavelier
CEA-LETI, 17 rue des Martyrs, 38054 Grenoble, France

B. Ghyselen
Soitec, Parc Technologique des Fontaines, 38926 Crolles Cedex, France

ABSTRACT

Compressive strained Silicon from a Silicon on Insulator (SOI) substrate is obtained by replacing the buried oxide layer by a strained silicon nitride layer. The silicon overlayer and the buried dielectric are etched down to the substrate to form narrow wires (down to 300 nm wide). The Si overlayer is then expected to acquire compressive strain thanks to the relaxation of the SiN layer. The goal is to obtain a high uniaxial stress perpendicular to the wires. The structures and the strain are modeled using finite element simulations. The strain elements are used to calculate Raman spectra. Theoretical results are compared to experimental profiles deduced from resonant (UV) micro Raman experiments.

INTRODUCTION

To insure expansion of semiconductor industry, the downscaling of CMOS technology remains the main strategy. For future technology nodes, electrical performance of transistors needs to be improved, for instance by carrier mobility increase. This can be achieved by several techniques that use local and/or global stress engineering in the transistor channel region. This includes strain modulation by using SiGe source and drain, or by depositing strained silicon nitride (SiN) layers over devices [1]. Alternative methods have been studied, such as wafer level introduction of strain, and in particular the use of Strained Silicon On Insulator substrates (Smart Cut[TM] technology applied to transfer of a strained Si layer made by heteroepitaxy on top of a relaxed SiGe template), which benefits from advantages of both SOI and stress engineering [2]. We propose to take advantage of the underneath region of the SOI stack to introduce strain into the active region, by introducing a strained silicon nitride (SiN) layer in the buried oxide (BOX) usually made of pure SiO_2. This technique of alter-BOX can be considered either as stand-alone or as an additional method to any of the above-mentioned techniques for channel strain engineering. For this purpose, advanced SOI substrates consisting in a 140 nm thick SiN layer covered by a ultra-thin Si film (about 10 nm) were fabricated. The initial SiN layer is under tensile strain. The silicon overlayer and the buried dielectric are etched down to the substrate to form narrow strips. The overlayer is then expected to acquire compressive strain thanks to the relaxation of the SiN layer. The goal is to obtain a high uniaxial stress perpendicular to the strips.

EXPERIMENTAL AND THEORETICAL DETAILS

Fabrication of SOI wafer with buried silicon nitride

The substrates with alternative buried dielectric layers are fabricated using the Bonded and Etched-back SOI (BESOI) technology (Fig. 1).

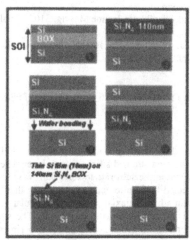

Figure 1: Schematic views of the Silicon On Insulator substrate fabrication with a silicon nitride buried stack.

A 140 nm silicon nitride layer is deposited at 780°C by Low Pressure Chemical Vapor Deposition (LPCVD) on a 200 mm SOI donor wafer, followed by annealing at 1100°C for densification (pictures 1 and 2 in Fig. 1). Direct chemical bonding (picture 3) without polishing process is allowed thanks to the very flat surface: RMS roughness values below 0.25 nm were measured using Atomic Force Microscopy on 140 nm as-deposited layers. Hydrophilic bonding is made between the SOI/Si$_3$N$_4$ donor wafer and a standard Si wafer. We found that a standard RCA cleaning protocol on both donor and base wafers is sufficient for hydrophilic nitride bonding [3]. The bonded substrates are then annealed at 1100°C in order to reinforce the bonding interface. The Si substrate and the SiO$_2$ BOX of the SOI donor wafer are then removed by grinding and chemical etching (pictures 4 and 5 in Fig.1).

The resulting structure (picture 6), consisting in a 10 nm thick Si film on the SiN buried layer, is etched down to the Si substrate to form narrow strips (few microns to 300 nm wide) oriented along [110] direction. In such condition, the tensile strain of the SiN layer is relaxed, inducing a compressive strain in the Si overlayer.

Computation of strain fields by finite element simulations

Finite element simulations were performed in order to compute the strain fields within the top silicon layer. The model is presented in the insert of Figure 2. It consists of half the narrow strip (because of symmetry) and also takes into account a large enough thickness of the silicon substrate to ensure that the numerical results are not affected by the boundary conditions:

Dirichlet conditions are imposed on the normal component of the displacement field at the bottom of the substrate and on the right and left sides of the model (see insert of Figure 2). The finite elements exhibit *large elastic strain* capability and the simulation is made under the *plane strain* hypothesis. This last condition is justified by the fact that the strip lengths are always large compared to the height of the etched stack. The nominal parameters used in the simulation are given for indication in Table I.

	Si	Si₃N₄	SiO₂
Young modulus (GPa)	130	370	74
Poisson ratio	0.28	0.28	0.17
Intrinsic stress (GPa)	--	1.0	--

Table I: nominal mechanical properties and intrinsic stress

Loading consists in an initial intrinsic biaxial stress state imposed in the buried nitride layer ($\sigma_{XX} = \sigma_{ZZ} = 1$ GPa). This stress/strain state is the mechanical equilibrium for full wafer deposit and is relaxed to the stress/strain state illustrated in Fig. 2 due to the etch step. Fig. 2 illustrates the σ_{XX} obtained in the thin silicon film for various strip widths and with elastic properties presented in Table I. This stress variation, corresponding to the direction where the carrier mobility has to be improved, should agree with the experimental results presented in the following. Contrary to the initial stress, measured by wafer bending, the Young's Modulus is not as easily determined and can vary in the 160-370 GPa range, depending on the references [4-5].

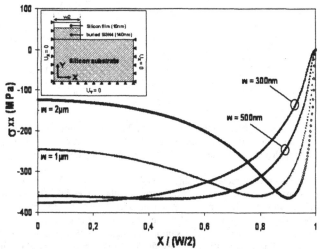

Figure 2: σ_{XX} stress fields extracted at mid-thickness for strip width of 300, 500, 1000 and 2000 nm; insert: model with boundary conditions. Raman spectra are recorded along the X axis.

Calculated and experimental Raman spectra

In absence of any applied stress, the zone center Si optical phonon is triply degenerated (located at 522 cm⁻¹) and only the component polarized along Y axis is Raman active. Stress lifts the degeneracy and may activate other components. In standard cases such as biaxial or uniaxial stress, it is easy to obtain a strain or stress value from the stress-induced shift of the Si phonon peak. In case of more complex geometries, and to describe singularities (line edges for instance), it is necessary to model Raman spectra [6-8].

In each point in the XY plane of the FE model we calculate a Raman spectrum: (i) The strain-induced frequency shifts and eigenvector changes are deduced from the strain tensor, by solving the usual secular equation (see equation 4 in Ref. 6), (ii) The Raman activities of the three optical phonon modes are calculated, considering the Raman tensors, the new eigenvectors, and light absorption [6].

Once all the Raman spectra have been calculated along the X axis, these spectra are convoluted with a Gaussian function taking account of the spatial resolution due to finite laser spot size. The so-obtained calculated spectra are then compared to the experimental ones. Until the agreement, with respect to the phonon peak shifts and intensities, is satisfactory a parameter in the finite element simulations can be adjusted and the procedure done once again.

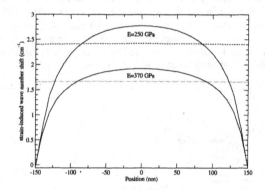

Figure 3: Calculated strain-induced Raman shifts in a 300 nm wide strip for different silicon nitride Young's Modulus values (full lines). Dashed and dotted lines correspond to the profiles convoluted with a Gaussian function taking account of the finite laser spot size.

The Raman spectra are recorded using an UV-dedicated Dilor XY spectrometer. Linear scans (100 nm step) are made along a line perpendicular to etched strips. Two to three acquisitions of about 15 sec each were used. Resonant excitation at $\lambda = 364$ nm is used in order to limit the probe depth (about 10 nm, equal to the strained layer thickness) and to enhance the signal. The laser beam is focused on the sample through a 0.95 Numerical Aperture (NA) microscope objective. The transmission is only 50% at 364 nm but it allows a high spatial resolution. Theoretically, the lateral resolution is given by $0.61/NA \approx 235$ nm, which is the value taken into account for the Gaussian function used for convolution. However, the main limitation is the 100 nm step of the displacement table, which makes difficult positioning the laser spot and limits the number of measurements on the narrowest strips. Another experimental issue, due to both the strained Si layer reduced thickness and dimensionality, is the sample heating under laser irradiation, which is evidenced by both the red shift and Full Width at Half Maximum (FWMH) enlargement of the phonon peak. Optimized operating conditions are obtained when the FWHM on the Si nanostructure is the same as for bulk Si (no confinement effect has to be taken into account for the sizes of interest in this paper). Eliminating local heating, the phonon peak blue shift compared to the substrate can be fully attributed to compressive strain

According to computed strain field values, we find similar behavior with increased strain-induced Raman shift near strip edges than in the center for wide strips. We focus however on the 300 nm wide strip since it the most interesting feature for applications (high and homogeneous stress σ_{XX}). The experimental wave number profile across a 300 nm wide strip is shown in Fig. 4.

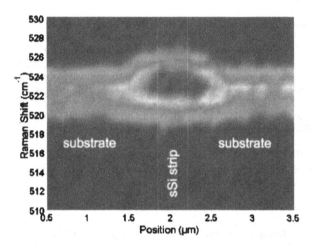

Figure 4: Experimental Raman spectra recorded in each point upon scanning across a 300 nm wide strip (the center of the strip is at about 2 μm). The color scale corresponds to the phonon intensity in the spectra (from blue, low intensity, to red, high intensity). The high intensity region corresponds to the simultaneous presence of both the substrate and strained Si phonon peaks.

DISCUSSION

In Fig. 4, the blue shift observed in spectra taken near or on the strip, compared to the substrate wave number on both sides, proves that compressive strain has been transferred from the SiN alter-box to the Si overlayer.

We investigated different fitting methods to extract the strain-induced shift in the strip, in particular to account of the substrate contribution since the laser spot is larger than the 300 nm wide strip. It is obvious looking at Fig. 4, in which the region exhibiting a blue-shit extends over about 0.8 μm. To fit the spectra recorded on or near such strained strip, two phonon peaks were needed. As a matter of fact, both the FWHM and intensity increase for spectra containing two contributions (strip and substrate).

In such conditions, the experimental strain-induced shift is 1.5 ± 0.1 cm^{-1}. This would correspond to about -0.78 GPa in the model of pure uniaxial stress [7], which is not the case here. Actually, it is close to the value obtained in Fig. 3 for the convoluted profile corresponding to SiN having a Young's Modulus of 370 GPa. The compressive stress σ_{xx} along the (110) direction is therefore close to -0.4 GPa (Fig. 2). It is rather strange that the experimental value corresponds to the highest Young's Modulus. Decreasing the Young's modulus in the computation of strain fields would however increase the transmission of strain to the Si layer, thus increasing both the strain-induced Raman shift and the discrepancy between experiment and calculation. Higher strain transmission could be achieved by increasing the tensile biaxial stress in the silicon nitride layer, while decreasing its Young's Modulus.

CONCLUSIONS

We have shown that compressive strain silicon lines could be obtained by processing SOI wafers in which the buried oxide was substituted by a tensile strained silicon nitride layer. The strain was modeled by finite element simulations, and subsequently used for the calculation of Raman spectra. The strain-induced Raman shifts were compared to experimental spectroscopic images. For thin (10 nm) and narrow (300 nm) Si wires, the compressive stress is quite uniform and reached several hundreds of MPa. Experimental issues are limited spatial and spectral resolutions and laser heating of submicron wide Si wires. Further works are planned to improve materials quality (mechanically and electrically), characterization of mechanical properties, and experimental conditions to measure accurate stress profiles.

ACKNOWLEDGMENTS

This work is partly supported by French National Research Agency (EXTRADA PNANO program ANR-06-NANO-042-06).

REFERENCES

1. Yee-Chia Yeo, Jisong Sun, Eng Hong Ong, *Mat. Res. Soc. Symp. Proc.* Vol. **809**, B10.4.1 (2004), and refs. therein.
2. B. Ghyselen *et al.*, *Solid State Electronics* **48**, 1285 (2004).
3. O. Rayssac, H. Moriceau, M. Olivier, I. Stoemenos, A. M. Cartier, B. Aspar, From SOI to SOIM technology: Application for specific semiconductor processes. Proc. 10th Int. Symp. on Silicon-on-Insulator Technology and Devices, PV 2001-3, 39. Pennington, NJ: *Electrochem. Soc. Proc. Series* (2001).
4. A. Khan, J. Philip, P.Hess, *J. Appl. Phys.* **95**, 1667 (2004) and refs. therein.
5. T. Yoshika, T. Ando, M. Shikida, K. Sato, *Sensors and Actuators* **82**, 291 (2000).
6. I. De Wolf, H. E. Maes, S. K. Jones, *J. Appl. Phys.* **79**, 7148 (1996)
7. I. De Wolf, M. Ignat, G. Pozza, L. Maniguet, H. E. Maes, *J. Appl. Phys.* **85**, 6477 (1999).
8. E. Latu-Romain, M. Mermoux, A. Crisci, D. Delille, L. F. Tz. Kwakman, *J. Appl. Phys.* **102**, 103506 (2007).

Mater. Res. Soc. Symp. Proc. Vol. 1185 © 2009 Materials Research Society   1185-II05-02

Geometrical Critical Thickness Theory for the Size Effect at the Initiation of Plasticity

T.T. Zhu, B. Ehrler, A.J. Bushby and D.J. Dunstan
Centre for Materials Research, Queen Mary, University of London, London E1 4NS, UK

ABSTRACT
Recently, size effects in the initiation of plasticity have been clearly observed and reported in different geometries; e.g., bending (Ehrler et al. Phil. Mag. 2008), twisting (Ehrler et al., MRS, Spring Meeting 2009) and indentation (Zhu et al. J. Mech. Phys. Sol. 56, 1170, 2008). Strain gradient plasticity theory is the principal approach for explaining size effects during plastic deformation in these geometries. However, it fails to account for any size effect at the initial yield. Geometrical critical thickness theory was proposed to explain the yield size effect in bending and torsion (Dunstan and Bushby, Proc. Roy. Soc. A460, 2781, 2004). The theory shows that the initial yield strength is scaled with the inverse square root of the characteristic length scale without requiring any free fitting parameters. Here, we extend the theory to describe the yield size effect in indentation. The theory agrees fairly well with experimental observations in micro-torsion (Ehrler et al., MRS, Spring Meeting 2009) and nanoindentation (Zhu et al., J. Mech. Phys. Solid, 2008).

1. INTRODUCTION

It has been known for several decades that materials display strong size effects at the micron or sub-micron scale, in which the strength is enhanced when the size of the structure or of the stressed volume is decreased. Generally, size effects can be categorized as intrinsic and extrinsic size effect [1]. Intrinsic size effects are due to microstructural constraints of materials. Microstructural size effects include those due to grain boundaries [2] and particle reinforcement [3]. The extrinsic size effects have been presented for uniform deformation (without introducing plastic strain gradient) and non-uniform deformation. Uniform size effects are observed in, e.g., compression of pillars [4] and tension of whiskers [5]. Non-uniform or strain-gradient size effects are seen in torsion of wires [6, 7], bending of foils [8, 9], indentation [10-12], etc.

Yield is always an important phenomenon but difficult to measure in materials science and engineering. Recently, size effects in the initiation of plasticity have been clearly observed and reported in different geometries, e.g, twisting [7] and indentation [12]. In all these different geometries, the yield strength clearly increases as the dimensional length scale is decreased.

Various theories account for the size effects. Their application is often controversial, and each may account for the size effect in a different situation, i.e., different strain regime (flow strength or yield strength, gradient or uniform). Strain-gradient plasticity theory is well-known, in which the size effect is attributed to hardening due to geometrically-necessary dislocations (GNDs) [13, 14]. Since GNDs can only exist in the non-uniform plastic deformation (Ashby, 1970), strain gradient plasticity theory is not able to explain the initial yield size effect. Slip-distance theory is based on the ideas of Conrad et al. [15], and has been more widely applied recently [16, 17]; in this theory the strengthening of materials is due to the limitations on dislocation mean free path. This theory can explain the flow strength size effect in uniform (without strain gradient) and non-uniform deformation. However, it is not able to explain the initial yield size effect. Critical thickness theory is a rigorous thermodynamic theory proposed initially for metals in 1949 [18] and developed since then primarily for semiconductors [19, 20]. The yield stress is predicted to follow an inverse dependence on the thickness h for a uniformly strained epitaxial crystal layer. Recently, critical thickness theory has been simplified, to account for the dimensional size effect in soft metals, as geometrical critical thickness theory [20-22].

Here, we describe the geometrical critical-thickness theory to the yield size effect in torsion and indentation. The theory shows that the initial yield strength is scaled with the inverse square root of the characteristic length scale without requiring any free fitting parameters. The theory agrees fairly well with experimental observations in micro-twisting [7] and indentation [12].

2. GEOMETRICAL CRITICAL THICKNESS

The underlying idea of the geometrical critical thickness theory is that generation of the dislocation is a cooperative process involving many atoms in the crystal and this process necessarily involves a finite volume rather than beginning at a point [20, 21]. This geometrically required small deformation area or volume restricts the generation (or movement) of dislocations and hence a higher yield stress is required. The additional (geometrically required) yield strain:

$$\Delta\varepsilon \approx \frac{b}{h_c} \qquad (1)$$

where h_c is the critical thickness for initial yielding and b is the effective Burgers vector of the dislocations. In this model, if dislocation multiplication (significant relaxation) is considered, b in Eq. (1) is to be replaced by ~$5b$ [20-22]. Equation (1) gives what we call the geometrical strength, or the geometrical contribution to the yield strain, of a thin strained layer of thickness h. For a strained semiconductor layer at growth temperature, the geometrical strength is often the only significant determinant of plastic yield. In the experiments on metals that we consider here, the elastic strains are much smaller and the bulk yield strain ε_Y, at which the metal would yield if the layer were very thick, is also relevant. Then the total yield strain is expressed as:

$$\varepsilon'_Y = \varepsilon_Y + \Delta\varepsilon = \varepsilon_Y + \beta\frac{b}{h_c} \qquad (2)$$

where ε_Y is the yield strain for bulk material. β is a coefficient varied between 1 to 10, depending on the operation of dislocation sources, and also on crystallographic trigonometrical factors (Schmid factor etc) causing the effective Burgers vector to be less than the b). Details are in [22].

3. GEOMETRICAL CRICTICAL THICKNESS IN TWISTING A THIN WIRE

When a length L of wire of radius R is twisted through an angle φ and the wire is fully elastic, the maximum elastic stress is at the surface. Up to yield, the surface stress τ_Y and the normalized torque T are given by:

$$\begin{cases} \tau_Y = \mu\kappa_Y R \\ T^{(N)}_Y = \dfrac{T_Y}{R^3} = \dfrac{1}{2}\pi\tau_Y \end{cases} \qquad (3)$$

where μ is shear modulus; κ_Y is the twist per unit length at yield.

However, from the geometrical critical thickness theory [20-22], a surface layer of thickness h at the onset of relaxation must be considered. This surface layer must yield cooperatively, so that it will contain no geometrically necessary (misfit) dislocations but will be separated by a layer of misfit dislocations from the elastically strained core. In simple epitaxial strained layers of semiconductor [21], the actual thickness h_0 of the surface layer is well defined and is determined by the growth specification. Here, it is a variable. The torque at yield varies with h, diverging to infinity at $h = 0$ and at $h = R$, and so h_0 is determined by the minimum of the torque at yield as a function of h.

Assuming that h is far smaller than the diameter of the wire, we can approximate the surface as flat and directly apply Eq. (2). Then in torsion, the average strain at yield (with the contribution of critical thickness geometrical strain) should be:

40

$$\varepsilon'_Y = \frac{\tau_Y}{\mu} + \frac{\beta b}{h} \tag{4}$$

This is the average strain required in the surface layer. Since the elastic strain rises linearly with radius, the average value from $r = R - h$ to $r = R$ is also the actual value at radius $r = R - h/2$. The yield occurs at the twist κ'_Y:

$$\kappa'_Y (R - \frac{h}{2}) = \frac{\tau_Y}{\mu} + \frac{\beta b}{h} \tag{5}$$

It is now straightforward to calculate the torque at yield, as a function of h, and to differentiate it with respect to h to find the minimum torque at $h = h_0$. This calculation in [22] obtains:

$$h_0 = \frac{-\beta b + \sqrt{\beta b^2 + 2\beta \varepsilon_Y bR}}{\varepsilon_Y} \tag{6}$$

and for $h_0 \ll R$, Eq.(7) comes to:

$$h_0 = \sqrt{\frac{2\beta bR}{\varepsilon_Y}} \tag{7}$$

so that the yield stress for the wire due to critical thickness theory is expressed as:

$$\tau'_Y = \tau_Y + \mu\sqrt{2\beta b\varepsilon_Y / R} \tag{8}$$

From Eq. (8), the yield strength is increased inversely scaled with the square root of radius of the wire. Then, the normalized torque from Eq. (3) is:

$$T^{(N)}_Y = \frac{1}{2}\left(\tau_Y + \sqrt{\beta}\mu\sqrt{2b\varepsilon_Y / R}\right) \tag{9}$$

We have recently made studies of the twisting torque induced in polycrystalline copper wires as a function of curvature [7]. The experiments are based those reported by Fleck et al. [6] but using a load-unload technique and a much longer specimen for higher accuracy. The full data-set and details of the experimental methods is published elsewhere [7]. With the accurate measurement, the yield size effect is clearly observed, where the yield strength is increased drastically with thinner wires.

In order to compare the theory with the experimental data in [7], where the grain size and wire diameter are both contributing to the yield size effect, we introduce an effective length l_{eff}. The grain size and wire diameter are considered to contribute symmetrically (as in [23, 17]), so,

$$\frac{1}{l_{eff}} = \frac{1}{R} + \frac{1}{d} \tag{10}$$

Then substituting the R in Eq. (9) by l_{eff} as in Eq. (10), the torque comes to,

$$T^{(N)}_Y = \frac{1}{2}\pi\left(\tau_Y + \mu\sqrt{2b\varepsilon_Y\left(\frac{1}{R} + \frac{1}{d}\right)}\right) \tag{11}$$

The theoretical fit to one datum for 10μm diameter Cu wire and three data for 50μm diameter Cu wire is obtained from Eq. (12), with parameter values of $\tau_Y = 3.5$MPa, $\mu = 48.3$GPa and $b = 0.256$nm and with β as the only fitting variable. A best fit is obtained by applying $\beta = 10$. The result is plotted as the dotted line on Fig. 1. It can be seen that the fitted strength with wire radius and grain size is consistent with the measurements.

41

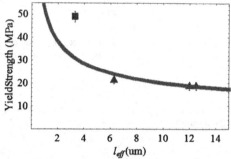

Figure 1 Polycrystalline copper twisting data obtained by Ehrler et al. [7]. Yield strength is plotted against the effective length l_{eff}. Cubic point is for: $R = 5\mu m$; Triangle point: $R = 25\mu m$. The solid lines are predicted from critical thickness theory as in Eq. (12), with $\beta = 10$ and $\tau_Y = 3.5$MPa.

4. GEOMETRICAL CRITICAL THICKNSS IN INDENTATION

Now we extend the theory to the indentation yield size effect. Here, we consider the indentation stress-strain field by using the 2-D cavity expansion model [24] as illustrated in Fig.2.

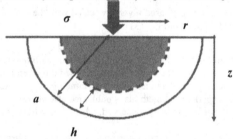

Figure 2. Cylindrical indentation with contact radius a, under a stress σ. The impression zone can be assumed as a hemisphere with radius a.

Assuming that with the applied stress σ, the indenter moves through a displacement u_0. The displacement (at $r = 0$), along the z direction can be approximately expressed as:

$$u_z = \frac{a}{z} u_0 \tag{12}$$

where a is the radius of the cavity and also the contact radius of the indent. The strain can be approximately expressed as:

$$\varepsilon_z = \frac{u_z}{z} = \frac{a u_0}{z^2} \tag{13}$$

The yield stress can be expressed as:

$$\sigma_y = \frac{E a u_Y}{a^2} = E \frac{u_Y}{a} \tag{14}$$

where E is the Young's modulus.

However, considering the geometrical critical thickness phenomenon, there is a thin elastic layer with thickness h (illustrated in Fig. 2) at the onset of relaxation, and this surface layer must yield cooperatively with the cavity. This corresponds to the yield strain $\varepsilon_Y + b / h$. Now we calculate the average strain from $z = a - h$ to $z = a$, and find that the average strain happens at z:

42

$$z = \frac{1}{h} \int_{a-h}^{a} \frac{1}{z^2} dz = \frac{1}{h}(\frac{1}{a-h} - \frac{1}{a}) \tag{15}$$

So the yield happens when displacement u_Y is:

$$\varepsilon'_Y = \frac{au_Y}{z^2} = \varepsilon_Y + \frac{\beta b}{h} \tag{16}$$

Solving Eq.(16) for u_Y,

$$u_Y = \frac{1}{a}(\varepsilon_Y + \frac{b}{h})\left[\frac{1}{h}\left(\frac{1}{a-h} - \frac{1}{a}\right)\right]^2 \tag{17}$$

Then the yield stress due to the critical thickness is:

$$\sigma_Y = \frac{Eu_Y}{a} = u_Y = \frac{E}{a^2}(\varepsilon_Y + \frac{\beta b}{h})\left[\frac{1}{h}\left(\frac{1}{a-h} - \frac{1}{a}\right)\right]^2 \tag{18}$$

We solve h_0 by finding the smallest yield stress σ_Y. Differentiating Eq. (18) with respect to h, setting the differential to be 0, then:

$$h_0 = \frac{-3\beta b + \sqrt{9\beta b^2 + 8\beta ba\varepsilon_Y}}{4\varepsilon_Y} \tag{19}$$

Since $h_0 \ll a$,

$$h_0 = \sqrt{0.5\beta b a / \varepsilon_y} \tag{20}$$

$$\sigma'_Y = \sigma_Y + \sqrt{\beta}E\sqrt{\frac{b\varepsilon_Y}{2a}} \tag{21}$$

Recently, Zhu et al. [12] measured the yield strength in spherical indentation of a series of ceramics. The results are plotted as in Fig.3. In order to compare this expression with experimental indentation data, we evaluate the indentation mean pressure P_m as [25],

$$P_m = 3\sigma_m \tag{22}$$

The theoretical prediction of P_m is therefore,

$$P'_Y = 3(\sigma_Y + \sqrt{\beta}E\sqrt{\frac{b\varepsilon_Y}{a}}) \tag{23}$$

Using Eq. (23), with parameter values listed in Table 1. β is the only free parameter. The best fit is with $\beta = 7.5$. The results are plotted as the solid lines in Fig 3. The predicted variation of strength with contact radius is highly consistent with the measurements.

It is interesting that these data are also well fitted by slip-distance theory [16]. In both critical thickness theory and slip distance theory the size effect depends only on the Burge's vector b, bulk yield strain ε_Y, shear modulus μ and a dimensionless fitting parameter on an order of unity.

Table1. Properties of the materials studied.

	Orientation	Young's modulus E / GPa	Burgers vector b / nm	Macroscopic yield stress σ_Y/GPa
α-Al$_2$O$_3$	$(11\bar{2}0)$	420	0.475	4.6
InP	(100)	91	0.414	0.67
In$_{0.53}$Ga$_{0.47}$As	(100)	93.8	0.413	1.3
GaSb	(100)	86.5	0.430	0.90

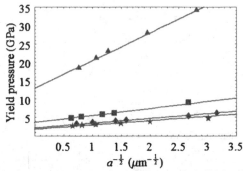

$$a^{-\frac{1}{2}} \ (\mu m^{-\frac{1}{2}})$$

Figure 3 Spherical yield pressure plotted for α-Al_2O_3(▲), $In_{0.53}Ga_{0.47}As$(■), GaSb(♦) and InP(★) as function of $a^{-1/2}$ [12]. Solid lines are the fitting by geometrical critical thickness theory as in Eq. (23), with $\beta = 7.5 \pm 0.1$.

5. CONCLUSIONS

The geometrical critical thickness theory is extended to yield size effect in twisting and indentation. It shows that the initial yield strength is scaled with the inverse square root of characteristic length scale. The theory agrees well with experimental observations in micro-twisting [7] and indentation [12].

REFERENCES

1. T.T. Zhu, A.J. Bushby and D.J. Dunstan, *Mater. Tech.* **23** (4), 193 (2008).
2. E.O. Hall, *Proc. Phys. Soc. Lond. B*, **64** (9), 747 (1951)
3. D. J. Lloyd, *International Materials Reviews* **39** (1), 1 (1994)
4. J.R. Greer, W.C. Oliver and W.D. Nix, *Acta Mater* **53**, 1821 (2005)
5. S.S. Brenner, *Science*, **128**, 568 (1956)
6. N.A. Fleck, G.M. Muller, M.F. Ashby and J.W. Hutchinson, *Acta Met.* **42**(2), 475 (1994)
7. B. Ehrler et al., *Phys. Rev. Letter.* submitted (2009)
8. J. S. Stölken, and A. G. Evans, *Acta Metallurgica* **46**(14), 5109 (1998)
9. B. Ehrler, R. Bossis, S. Joly, K. P'ng, A. Bushby and D. Dunstan, Phys. Rev. Letter., submitted.
10. Q. Ma and D.R. Clark, *J. Mater. Res.* **10**, 853 (1995)
11. J.G. Swadener, E.P. George, and G.M. Pharr, *J. Mech. Phys. Solids* **50**, 681 (2002)
12. T.T. Zhu, A.J. Bushby and D.J. Dunstan, *J. Mech. Phys. Solids* **56**, 1170 (2008)
13. M.F. Ashby, *Phil. Mag.* **21**(170), 399 (1970)
14. Y. Huang, S. Qu, K.C. Hwang, M. Li, and H. Gao, *Int. J. Plasticity* **20**(4), 753 (2004)
15. H. Conrad, S. Feuerstein and L. Rice, *Mater. Sci. Eng.* **2**, 157 (1967)
16. A.J. Bushby, T.T. Zhu and D.J. Dunstan, *J. Mater. Res.*, **24**, 966 (2009)
17. T.T. Zhu, A.J. Bushby and D.J. Dunstan, J. Mech. Phys. Solids., submitted
18. F.C. Frank and J. H.ven der Merwe, *Proc Roy Soc (London)*: **A198**, 216 (1949)
19. J. W. Matthews, *Philos. Mag.*, 1966, **13**, 1207 (1974)
20 D.J. Dunstan, S. Young and R.H. Dixon, J.Appl.Phys., 70, 3038 (1991)
21. D.J. Dunstan: *Journal of Materials Science: Materials in Electronics*, **8**, 337 (1997)
22. D.J. Dunstan, and A.J. Bushby, *J. Proc. R. Soc. Lond. A*, **460**, 2781, (2004).
23. A. W. Thompson, *Acta Met.*, **23**, 1337 (1973)
24. R. Hill, The mathematical theory of plasticity, Oxford at the Clarendon Press., (1950).
25. D. Tabor: The Hardness of Metals. Clarendon Press, Oxford, (1951)

Mater. Res. Soc. Symp. Proc. Vol. 1185 © 2009 Materials Research Society   1185-II05-04

Computational Nanomechanics of Graphene Membranes

Romain Perriot, Xiang Gu and Ivan I. Oleynik
Materials Simulation Lab, University of South Florida, Department of Physics,
4202 East Fowler Avenue,Tampa, Fl 33620, U.S.A.

ABSTRACT

Molecular Dynamics (MD) simulations of nanoindentation on graphene membranes were performed. The 2-d Young's modulus of the graphene monolayer was determined as 243 ± 18 N/m and the breaking strength as 41 ± 3 N/m. These values agree reasonably well with recent experimental results [1]. In addition, the simulations allowed us to examine the atomic-scale dynamics of membrane breaking during the nanoindentation, involving formation of an increasing number of lattice defects until membrane is completely broken. The onset of defect appearance allowed us to determine the true elastic limit of graphene and the corresponding yield strength 29 ± 1 N/m which was not accessible experimentally. The defects consist of vacancies and Stone-Wales type defects. Long stable linear chains of sp bonded carbon atoms (carbynes) were observed under the indenter at the advanced stages of indentation. The dynamics of fracture propagation is governed by the shear stresses developed in the sample.

INTRODUCTION

Graphene is a carbon-based material consisting of a monolayer of covalently bonded carbon atoms arranged in a honeycomb lattice. Since its successful isolation from graphite, the material has drawn attention from the scientific community, due to its remarkable fundamental electronic, optical and magnetic properties as well as its promising applications in nanoelectronic devices [2]. Graphene also exhibits unusual mechanical properties. In particular, Lee et al. [1] reported results of nanoindentation experiments on graphene membranes by an atomic force microscope (AFM) tip. It was found that it has an exceptional breaking strength, making it the strongest material studied so far. This opens up exciting opportunities for mechanical applications of graphene ranging from resonators and pressure sensors to carbon-fiber reinforcements.

The goal of this work is to perform computational experiments of nanoindentation of graphene membranes using atomic-scale simulation techniques. Massively-parallel molecular dynamics (MD) simulations allowed us to extend the size of the system, approaching micron-size. More importantly, the mechanical properties of graphene membranes were studied to a level of detail that is difficult or sometimes impossible to obtain in experiment. In particular, the appearance of defects in the course of indentation enabled us to determine the true elastic limit and the corresponding yield strength which was not accessible experimentally.

Two subsets of membranes were employed in our simulations. A first subset of small-diameter membranes was used to identify the most interesting features, thus saving computational time. Once the important physics and regimes were found, they were thoroughly investigated using a large subset of membranes with diameter approaching experimental dimensions. The fundamental mechanical properties of graphene membranes were predicted, namely the 2-d Young's modulus, the yield and the breaking strengths. In addition, defect formation mechanisms and breaking dynamics were investigated with atomic-scale resolution.

COMPUTATIONAL DETAILS

The simulations were performed with the MD code LAMMPS (Large-scale Atomic/Molecular Massively Parallel Simulator, [3]). The carbon-carbon interactions were described by the Reactive Empirical Bond-Order Potential (REBO) [4], which accurately reproduces the behavior of graphene and diamond structures near equilibrium. The spherical indenter was represented by a strong repulsive potential of the form:

$$V(r) = A(R - r)^3 \theta(R - r) \tag{1},$$

where A is a constant describing the strength of the interaction, R is the radius of the "indenter", and $\theta(r)$ is the Heaviside step function.

Circular membranes were used in simulations, such that a circular layer of several angstroms in thickness along the edge was clamped to mimic the conditions of the nanoindentation experiments. Membranes of several diameters were divided into two subsets: a "small" subset with diameters 400, 600 and 800 Å and indenter radii of 10, 20 and 30 Å, and a "large" subset with diameters 2000, 2500 and 3000 Å, and 50 and 100 Å radii indenters.

The first set of computational experiments involved static loading of the membrane by the indenter. For a given indentation depth, a membrane surface was deformed to take the shape that originally flat membranes acquire under point load applied at the center of the membrane. Such deformation profile is obtained as an analytic solution within elasticity theory [5]. In order to account for the finite size of the indenter and possible non-linear deformations, the membrane was allowed to relax under load of the spherical tip described by formula (1). Then, the system was subjected to a temperature of 3000 K (which is well below the melting temperature for graphene) by running NVT MD simulations for 10 ps. Such step was necessary to activate defect formation at sufficiently large loads. The NVT activation step was followed by a gradual cooling to zero temperature and a final static relaxation after which the force exerted on the indenter was calculated.

By running a series of simulations at different indentation depths, the indentation curves (load force vs indentation depth) were obtained. In order to compare with experimental results by Lee et al., the simulated indentation curves were fitted to the expression for the elastic membrane under load and prestress σ_0 [6]:

$$F(\delta) = \sigma_0 \pi \delta + E^{2D} q^3 a (\delta/a)^3 \tag{2},$$

where δ is the indentation depth, E^{2D} is the 2-d Young's modulus, q is a parameter defined by Poisson's ratio (theoretical value – REBO [7]) and a is the diameter of the membrane.

If comparison with bulk materials is required, the 3-d Young's modulus can be obtained from the 2-d value E^{2D} and effective thickness of the 2-d material, which is rather ambiguous for graphene [7]. Therefore, the 2D Young's modulus E^{2D} was chosen as the most fundamental quantity of interest in the present work. According to the theory of elasticity [8], the yield strength σ^{yield} of the membrane, corresponding to the elastic limit is defined as:

$$\sigma^{yield} = \left(\frac{F^{yield} E^{2D}}{4\pi R} \right)^{1/2} \tag{3},$$

where R is the radius of the indenter, F^{yield} the load on the membrane corresponding to the onset of plastic deformations (or defect appearance).

A second computational experiment involved a dynamic indentation of the membrane. The indenter is moved towards the membrane at constant speed, by running NVT MD simulations at 300 K, until the membrane is completely broken. Due to appreciable computational expense, the indentation process was started from an initial geometry chosen from the static indentation simulation which is still well below the elastic limit. During the dynamical indentation, the atomically-resolved stresses and the potential energy were recorded for subsequent analysis.

RESULTS AND DISCUSSION

Static loading

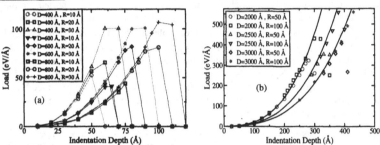

Figure 1: Indentation curve (force versus indentation depth) for the two membrane subsets: small diameter membranes (a) and large diameter membranes (b). D is the membrane diameter and R is the indenter radius.

The indentation curves obtained from computational nanoindentation experiments are shown in Fig. 1. As a function of indentation depth, each curve exhibits a similar behavior: a linear behavior of the load force at small indentations followed by a rapid non-linear increase until the appearance of defects. As the number of defects increases, the force exerted on the indenter increases slower, before reaching a plateau and finally decreasing to zero after complete breaking. Two interesting features were observed for the "small" subset of membranes (Fig. 1a). First, for the same membrane with different tip radii, the curves are almost identical up to the points where defects start to appear. This is the consequence of the fact that the tip diameter is small compared to the diameter of the membrane. The deformations are determined by the magnitude of the load only (point indenter approximation) rather than the finite contact area with the tip. This is why the theoretical indentation curve (2) is independent on the tip radius.

Second, the maximum load sustained by the membrane before it breaks (breaking strength) does not depend on the membrane size, but rather on the indenter radius. This is because the breaking strength is controlled by the applied local pressure (force per unit area) in the case where the indenter diameter is much smaller than the membrane diameter. This is also valid for the yield strength corresponding to the limit of elasticity, see formula (3). These two observations (same curve for a single membrane but different tip diameters, and same yield and breaking strength for the same tip but different membrane diameters), were also noticed in experiment [1].

The same trends were found to be characteristics of the "large" subset (Fig. 1b). An accurate sampling of indentation depths was performed for the membrane of diameter 2000 Å and indenter radii $R = 100$ Å and $R = 50$ Å. Due to a substantial computational cost, fewer points were calculated for the two other membranes (2500 and 3000 Å). This was done mostly to validate the previous results extended to the samples of larger size approaching experimental dimensions. However, some deviations were observed regarding the breaking strength, which might be due to the sufficiently large tip size compared to the membrane diameter. This might modify the equation (3) used to extract the breaking strength.

Equation (2) was used to fit the curve for a membrane diameter of 2000 Å and tip radii $R = 100$ Å and $R = 50$ Å, using the 2-d Young's modulus and the prestress as fitting parameters. This yielded the following values: $E^{2D} = 243 \pm 18$ N/m, and $\sigma_0 = 0.2 \pm 0.04$ N/m. The indentation curves for the two other membranes of larger diameter were fitted using (2) producing results in good agreement with the values reported above. The value for the 2-d Young's modulus from our computational experiment is lower than that reported in the experiment [1], but in a good agreement with previous theoretical calculations [7], see Table I. Because we can clearly identify the elastic limit, we were able to obtain both the yield σ^{yield} and breaking σ^{max} strengths using equation (2) which was derived within elasticity theory [8]. We found a breaking strength in good agreement with experiment, see Table II. Taking into account that our simulations produced smaller 2-d Young modulus compared to experiment, this agreement seems surprising. It is possible though, that the quasi-static process of membrane loading leads to an overestimation of the maximum force sustained by the membrane, and thus an overestimation of the breaking strength.

Table I: 2-d Young's Modulus

	E^{2D} (N/m)
This work	243 ± 18
Experiment [1]	340 ± 50
Theory [7]	242 ± 1

Table II: Yield and breaking strengths

	σ^{yield} (N/m)	σ^{max} (N/m)
This work	29 ± 1	41 ± 3
Experiment [1]	-	42 ± 4

Figure 2 shows the formation of defects at several stages of the indentation. After holes are opened up in the region directly under the indenter, the structure is stabilized by long chains of *sp* bonded atoms. The existence of such structures (carbynes) has been discussed in the literature (see, e.g., ref. [9] for a classification of the different carbyne types), and they were observed in carbon films [10]. Even after complete breaking of the membrane, see Fig. 2d, stable chains can be seen hanging at the edges of the hole formed.

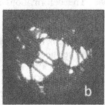

Figure 2: Central part of the membrane (D= 2000 Å, R= 100 Å) with snapshots corresponding to different indentation depths: a - 270 Å, b - 300 Å, c - 320 Å, and d - 350 Å. The linear size is about 250 Å, and coloring (online) reflects the coordination number of the atoms: green - coordination 3, beige - 2, turquoise - 1 and white - 0.

Dynamical Indentation

Dynamical nanoindentation experiments were performed using the 2000 Å diameter graphene membrane and 100 Å indenter radius. The dynamic indentation started from the initial indentation depth δ = 230 Å. The membrane was relaxed under static loading to prepare the initial configuration and the sample has been inspected to insure the absence of defects. Then, a constant velocity (2.5 Å/ps) was applied to the indenter until it reached an indentation depth of 350 Å. The MD indentation simulations were performed at a constant temperature of 300 K using the NVT ensemble. During the run, atomically-resolved stresses and potential energies of the atoms were recorded for further analysis. We found that the membrane breaking occurred at a smaller indentation depth compared to the static indentation experiments, thus exhibiting the importance of the dynamic effects during the course of defect formation in the membrane under the indenter. The rate of the indentation, 2.5 Å/ps, is fast as compared to experiment and this may also contribute to the deviation.

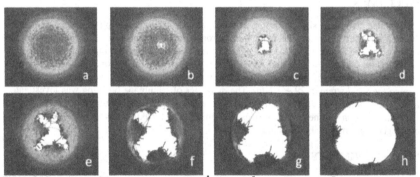

Figure 3: Central part of the membrane (2a= 2000 Å, R= 100 Å) during dynamic indentation, with snapshots corresponding to different indentation depths: a – 297; b - 297.5; c – 298; d - 298.5; e – 300; f – 305; g – 310; and h - 350 Å. The coloring depicts the atomically-resolved pressure (increasing from dark blue to red).

Figure 3 shows the pressure distribution under the indenter at several stages of indentation. We found that the loading pressure is concentrated in the area immediately under the indenter. As the local potential energy due to bond stretching reaches its critical value, the first defect appears, see Fig. 3b. This does not occur exactly at the center of the membrane due to the statistical nature of the process and the presence of thermal fluctuations. Once the localized defect appears, the fracture rapidly grows at the very small advance of the indenter (1 Å) by releasing the elastic energy and pressure by breaking the graphene bonds, see Fig. 3c, and 3d. The very localized distribution of energy and pressure within a rather small membrane/tip contact area explains both experimental and computational observations that the breaking strength does not depend on the membrane diameter. As long as the membrane diameter is large compared to the tip radius, the only atoms affected by the indentation are those that are under the indenter. The fracture continues to propagate until complete breaking of the membrane, see progressive snapshots e, f, g, h in Fig. 3.

Figure 4: A: Atomically-resolved shear stress distribution in the membrane under tip load. The regions of high magnitude shear stress are colored in dark blue (negative) and red (positive). B: Corresponding broken membrane geometry. Coloring is according to the atomically-resolved potential energy.

We found that the geometry of the broken membrane can be predicted by examining the shear stress distribution in the membrane before it breaks, see Fig. 4-A. It can be shown that the symmetry of the stress-tensor in case of 2-d systems results in localization of the shear stresses along two orthogonal directions. This recognizable "cross-pattern" distribution is clearly seen in Fig. 4-A which shows the calculated distribution of the shear stresses in the membrane before it breaks. Therefore, the shear stress distribution determines the resulting geometry of the damage.

CONCLUSIONS

Molecular Dynamics simulations of nanoindentation of circular graphene membranes were performed in order to obtain the mechanical properties of graphene and mechanisms of the breaking under the spherical tip load. The 2-d Young's modulus and the breaking strength were predicted to be 243 ± 18 N/m and 41 ± 3 N/m respectively, in reasonable agreement with experiment. The onset of plastic deformation was clearly identified, which allowed us to determine the yield strength 29 ± 1 N/m. The dynamics of membrane breaking was investigated and the formation of 1-d carbyne chains was observed during the indentation process. We found that the distribution of the shear stresses determines the breaking mechanism including the resulting geometry of membrane damage.

ACKNOWLEDGMENTS

This work is supported by the National Science Foundation (Grant No. 0726842). Calculations were performed using NSF TeraGrid facilities (Grants TG-DRM070018N and TG-MCA08X040), USF Research Computing Cluster, and computational facilities of Materials Simulation Laboratory at the University of South Florida funded by ARO (grant No W911NF-07-1-0212).

REFERENCES

[1] Lee et al., Science **321**, 385 (2008)
[2] Geim, A.K., Novoselov, K.S., Nature Materials **6**, 183-191 (2007)
[3] S. J. Plimpton, J Comp Phys, 117, 1-19 (1995), http://lammps.sandia.gov
[4] D. W. Brenner, Phys. Rev. **B 42**, 9458 (1990), J. Phys.: Condens. Matter. **14**, 783 (2002)
[5] S. Timoshenko, Theory of Plates and Shells, 2nd edition, Mc Graw-Hill (1959)
[6] U. Komaragiri, M. R. Begley, J. Appl. Mech. **72**, 203 (2005)
[7] Huang et al., PRB **74**, 245413 (2006)
[8] N. M. Bhatia, W. Nachbar, Int. J. Non-Linear Mech **3**, 307 (1968)
[9] R. B. Heimann et al., Nature **306**, 164 (1983)
[10] L. Ravagnan et al., Phys. Rev. Lett. **89**, 285506 (2002)

Mater. Res. Soc. Symp. Proc. Vol. 1185 © 2009 Materials Research Society   1185-II05-07

On the Effects of Dislocation Density on Micropillar Strength

A. A. Benzerga

Texas A&M University, College Station, TX 77843, USA

ABSTRACT

There is an increasing amount of experimental evidence that the plastic behavior of crystals changes at micro- and nano-scales in a way that is not necessarily captured by state-of-the-art plasticity models. In this paper, length scale effects in the plasticity of crystals are analyzed by means of direct numerical simulations that resolve the scale of the carriers of plasticity, i.e., the dislocations. A computationally efficient, atomistically informed dislocation dynamics framework which has the capability of reaching high dislocation densities and large strains at moderately low strain rates in finite volumes is recalled. Using this theoretical framework, a new type of size effect in the hardening of crystals subject to nominally uniform compression is discovered. In light of such findings, behavior transitions in the space of meaningful structural parameters, from forest-hardening dominated regime to an exhaustion hardening dominated regime are discussed. Various scalings of the flow stress with crystal size emerge in the simulations, which are compared with recent experimental data on micro- and nano-pillars.

INTRODUCTION

Over the past few years, various experimental techniques have been developed to interrogate the mechanical response of materials at the scale of their microstructures. Among these, compression pillars have been extensively used [1–4]. In general, a common trend emerges from pillar compression experiments with smaller being harder. However, there are conflicting reports on whether hardening is size-dependent, and if so what is the origin of the apparent hardening? In addition, the strength of the scaling of flow stress with pillar diameter varies from one experimental data set to another. Therefore, there is a need for further analysis of plasticity in small volumes, especially in the absence of gradients in the macroscopic fields. Under such circumstances, state-of-the-art strain-gradient and other nonlocal theories of plasticity do not predict the size dependence evidenced in pillar compression experiments.

Since continuum plasticity is currently incapable of providing a rationale for pillar plasticity, recourse to lower resolution analyses is necessary. Fully discrete atomic-level methods, such as molecular dynamics, are incapable of resolving sample sizes ranging from 100 nm to over 10 microns, i.e., the range of pillar diameters considered in the experiments thus far. Alternatively, semi-discrete analyses may be used which are based on dislocation theory, i.e., linear elasticity for long-range dislocation interactions as well as suitably specified atomic-level input. Discrete dislocation dynamics (DD) simulations have recently been reported [5–9] which capture various aspects of plastic behavior in micro- and nano-pillars. Despite this progress, it is fair to say that our understanding of the basic mechanisms and how these relate to macroscopic behavior is still incomplete. In particular, microstructural parameters that affect the size scaling of flow stress

.remain largely unknown. The objective here is to examine behavior transitions associated with variations in the initial dislocation source density and how that affects the flow stress scaling of micropillars.

THEORY

The framework of mechanism-based discrete dislocation dynamics [10] is used in the simulations. Dislocation loops are modeled as connected pairs of edge dislocations of opposite sign and restricted to gliding in their slip plane. Initial dislocation sources are randomly distributed with fixed density ρ_0. To maintain a near-zero net Burgers vector in the initial state forest dislocations are used as "centers" of multiplication. Thus, a source is defined by a pair of opposite-sign dislocations pinned each by a forest dislocation. The operation of source I is taken to mimic the Frank–Read process with critical stress

$$\tau_{\mathrm{nuc}}^I = \beta \frac{\mu b}{S^I} \tag{1}$$

where β is a constant and S^I is the source length. For the source to emit a fresh dipole, it is required that the Peach–Koehler force acting on it attains or exceeds $\tau_{\mathrm{nuc}}^I b$ during a critical time, which represents the time it takes for the bowing segment in the actual 3D process to reach its critical configuration. Based on line tension approximations, the critical time is derived in closed form as [10, 11]

$$t_{\mathrm{nuc}}^I = \gamma \frac{S^I}{|\tau^I| b} \tag{2}$$

where γ is a constant with units of a drag factor, and τ^I is the current resolved shear stress, exclusive of the junction self-stress. Thus, the spatial locations and lengths of initial sources are random and follow from the random generation of the centers of multiplication. The source size distribution has a natural higher cutoff defined by specimen size, as well as a lower cutoff $S_{\min} = 320b$. The default size of the emitted dipole is taken to be κS^I. Estimates of the constants β, γ and κ are used based on published 3D simulations, e.g., [12]; also [13]. Here $\beta = 5$, $\gamma = 1–10B$, where $B = 10^{-4}$ Pa s is the phonon drag factor, and $\kappa = 2$. Note that the random distribution of multiplication "centers" results through (1) and (2) in a random and wide distribution of initial source strengths and nucleation times.

A superposition framework [14] is used to account for the long-range elastic interactions among dislocations. The Peach-Koehler glide force acting on dislocation i is

$$f^i = \mathbf{m}^i \cdot \left[\hat{\sigma} + \sum_{j \neq i} \sigma^j \right] \cdot \mathbf{b}^i \tag{3}$$

where $\mathbf{m}^i$ is the slip plane normal, $\mathbf{b}^i$ the Burgers vector and σ^j the infinite medium stress field of dislocation j. Also, $\hat{\sigma}$ is the image stress computed based on the boundary-value problem solution using a finite element method. With reference to a face-centered cubic lattice, internal friction is omitted so that the dislocation velocity in free flight is given by $Bv^i = f^i + \mathcal{L}^i b^i$ with B as above and $\mathcal{L}^i$ the line tension [10].

The superposition procedure does not apply to dislocation cores where nonlinear effects are prominent. These are accounted for through a set of rules describing the short-range dislocation interactions. For instance, rules for dislocation mobility, as above, annihilation at a critical distance $L_e = 6b$ and glide out of the crystal are used [14, 15]. A key ingredient in the "2.5D" simulation paradigm consists of local rules to account for additional nonlinear core effects. These include rules for dislocation junctions resulting from short-range reactions, i.e., not mere attractive dipoles. Junction formation is taken to occur when non-coplanar mobile dislocations fall within a distance d^*. Based on arguments developed by Kubin and co-workers [16] regarding the importance of cross-slip in stabilizing junctions, we distinguish two populations. A junction can act as an anchoring point for a dynamic Frank–Read source, with probability p. It is postulated that p represents an effective cross-slip probability. As such, p is generally affected by strain rate, but weakly during stage II hardening, which is dominated by athermal forest-cutting processes. Here, we use $d^* = 6b$ and $p = 0.05$. Alternatively, the junction acts as a breakable obstacle. Atomistic simulations have elucidated one possible mechanism for junction destruction by unzipping [17]. Using a line tension approximation, the junction breaking stress is given by a formula similar to (1) with β replaced by a factor β' that reflects junction strength. Here $\beta' = 5$. The activation of a dynamic Frank-Read source follows exactly the same rules adopted for initial static sources with the additional requirement that during the "waiting time" (2), none of the pinning junctions breaks away. Junctions constitute the only obstacles in the simulations. A dislocation pinned at a junction is released only if the junction is destroyed.

Crystals having dimensions $H \times D$ are potentially oriented for double slip at $\pm 35.25°$ from the vertical compression axis. The width D is varied from 200 nm to a few μm while keeping the ratio H/D constant. A displacement-controlled loading is applied to the top boundary and traction-free conditions are imposed on the lateral surfaces. The nominal strain rate is the same, $10^3 s^{-1}$, for all specimen sizes. In presenting the overall responses, nominal measures of stress (σ) and strain (ε) are defined as the top-boundary average traction divided by specimen width D and the top-boundary displacement divided by specimen height H, respectively. Material parameters of Al are used with $\mu = 26$ GPa, $b = 0.25$ nm and $\nu = 0.3$ for Poisson's ratio.

RESULTS

Since much is known about bulk crystal plasticity, it would seem logical that a DD framework be first tested against it before simulation predictions can be examined for micro/nano pillar plasticity, which is only partially understood. The challenge is that the computational demands for predicting stress-strain behavior up to large strains and large dislocation densities along with the inherent pattern formation are still too high. In investigating size effects from the sub-micron to over ten microns and across wide variations of dislocation densities, "2.5D" simulations offer the potential of a unified explanation. The potential of such simulations has been demonstrated for predicting stage II hardening [10] and incipient pattern formation [15] and is illustrated by the results of Fig. 1. In the bulk limit, "2.5D" DD simulations deliver a hardening response, a scaling of the flow stress with the dislocation density which is consistent with the Taylor law and a refinement of the dislocation structure with increasing stress levels.

Quid of the size effect? It is within that same unifying framework that size effects are now investigated. In doing so, two cases are considered: a low dislocation source density ($\rho_0 = 10^{12} m^{-2}$) and a high dislocation source density ($\rho_0 \approx 10^{14} m^{-2}$). In each case the only independent parameter varied is the width D.

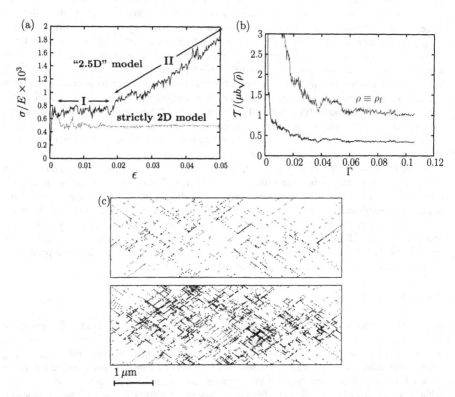

Figure 1: Results of 2.5D simulations in a "bulk" specimen with initial source density of $5 \, 10^{13} \mathrm{m}^{-2}$ (regime dominated by forest hardening) (a) Stress versus strain; (b) Flow stress (in units of $\mu b \sqrt{\rho}$) versus strain; and (c) Refinement of dislocation structure with accumulation of plastic strain. Also shown in (a) are results from a strictly 2D model (i.e., no junctions or dynamic sources) and, in (b), the scaling with the forest density.

In the low dislocation density case, a regime of exhaustion hardening [11] emerges (Fig. 2). Exhaustion hardening is an extended microplastic regime characterized by an imbalance between rates of dislocation multiplication versus escape at free surfaces. It is somewhat similar to the dislocation starvation picture depicted in [2] except that it is a dynamic concept. Consequently, the overall stress–strain curve changes drastically (Fig. 2a) and no net accumulation of dislocations occurs, Fig. 2b. In this regime, the average dislocation spacing is roughly of the order of specimen width for all sizes considered. Therefore, a stochastic response is expected and quantification of the flow stress is done in the ensemble average sense.

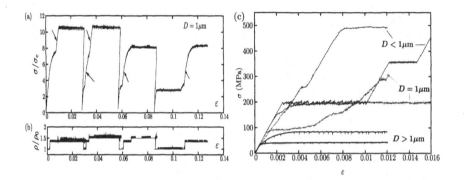

Figure 2: Results of 2.5D simulations in a specimen with initial source density of 10^{12}m^{-2} and aspect ratio of 6. (a) Stress versus strain ($\sigma_c = 25\text{MPa}$); (b) Total dislocation density versus strain. Also shown are a series of loadings and unloadings. Reloadings are purely elastic up to $\sigma/\sigma_c \approx 1$. Portions indicated by arrows have slopes lower than elastic and consist of many computational time steps. (c) Representative stress versus strain responses to compression of various specimens. The width D is varied while keeping the same initial dislocation density of 10^{12}m^{-2} (regime dominated by exhaustion hardening.)

Representative stress–strain responses of specimens of various sizes are plotted in Fig. 2c. The results are grouped into three sets: $D < 1\mu\text{m}$ (with lowest value 408nm); $D = 1\mu\text{m}$; and $D > 1\mu\text{m}$ (the largest width being 3266nm). $H/D = 6$ in all. There are two distinct, but related, contributions to this size effect. The first is the truncation induced by specimen size reduction in the distribution of initial source lengths, just as in previous DD analyses [5,8,9]. If realized in small crystals, short sources, which would be dormant in large crystals, require higher activation stresses. The second contribution comes from exhaustion hardening, which has subtle effects. In the smallest samples (typically containing one or two sources) a stair-like response is often obtained with no hardening during flow. In larger samples, significant (exhaustion) hardening occurs in between abrupt increases in the flow stress. The elastic or nearly-elastic segments indicate that low-probability events such as junction formation can have a drastic effect on the overall response in source-limited plasticity. The two contributions (source truncation and exhaustion hardening) are related, since both of them result from the scarcity of sources and the escape of dislocations at free surfaces.

By way of contrast, if the initial dislocation density is sufficiently large all crystals exhibit some steady strain-hardening, as illustrated in Fig. 3a. However, the analyses consistently predict a size effect on the hardening rate. The results of many such calculations are summarized in the plot of Fig. 3b giving the rate of rapid hardening versus specimen width D. For comparison, the data of Frick et al. [3] is also shown. Although the hardening mechanisms involved in the experiments would require careful analysis, it is gratifying to notice the excellent correspondence of hardening rates in the ensemble average sense.

In the high dislocation density regime, the average dislocation spacing is much smaller than specimen dimensions and any size effect must be related to the evolving dislocation structure. Following a method developed in [18] maps of GND densities were generated, an example of which is shown in Fig. 4a. An effective measure of GND density is then defined as the volume average of the GND density over the specimen (note that the net GND density essentially vanishes irrespective of size). Using such a definition, it was found that the effective GND density consistently increases with decreasing specimen size, Fig. 4b. This size effect is due to the emergence of deformation-induced dislocation

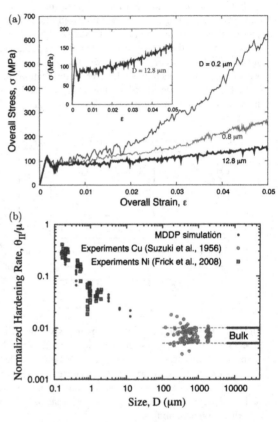

Figure 3: Results of 2.5D simulations in a specimen with initial source density of $1.5\,10^{14}\mathrm{m}^{-2}$ and aspect ratio of 3. (a) Sample stress versus strain curves; (b) Stage II hardening rate (in units of elastic shear modulus) versus specimen width D from simulations and experiments. MDDP stands for mechanism-based discrete dislocation plasticity. (regime dominated by forest hardening.)

structures with a characteristic length comparable with specimen dimensions. Guruprasad and Benzerga [7] explicitly characterized the evolving dislocation structures in terms of the effective density of geometrically necessary dislocations.

As analyzed by Guruprasad and Benzerga [7], the dislocation density increases at somewhat higher rates in smaller specimens. However, this trend alone is not enough to explain the additional strengthening predicted in the high-density calculations. Fig. 5 shows the comparison between the DD simulations and predictions from the Kocks-Mecking-Estrin (KME) dislocation-density based model [19]. In a sufficiently large crystal (i.e. one where the effective GND density is roughly zero, cf. Fig. 4) the DD simulations capture very well the hardening behavior predicted by the KME model (Fig. 5d). However, the smaller the specimen the larger the deviation from the KME prediction. It is worth emphasizing that integration of the KME equations was here carried out using the computed dislocation densities. The deviations would therefore be greater if the KME dislocation density evolution equation were directly used. Details may be found in Ref. [18].

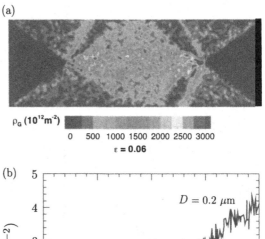

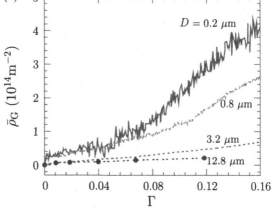

Figure 4: (a) Contour of GND density, ρ_G, at $\varepsilon = 0.06$ and a resolution of 50×50 nm^2 for crystal width $D3.2$ μm. (b) Effective GND density (integral measure over specimen) versus overall strain in four different specimens.

Fig. 6 shows the ensemble-averaged flow stress versus specimen size. Bars represent one standard deviation. In particular, if a power law is used to fit the results then the scaling exponent is found to be -0.83 in the low density case versus -0.46 in the high density case. Quantitatively, Fig. 6 shows that the scaling that results from forest-hardening related size effects is weaker than that associated with exhaustion hardening.

DISCUSSION

Plastic behavior in small volumes continues to pose challenges from the mechanics as well as physics perspectives. From the mechanics point of view, the mechanical response of a given ma-

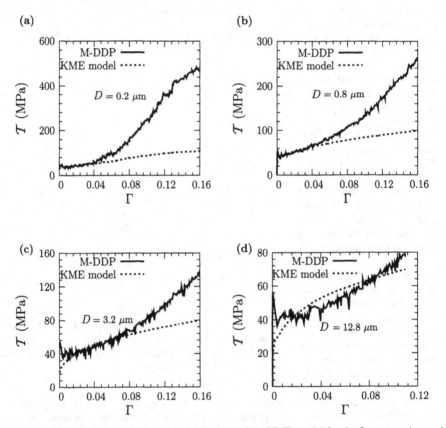

Figure 5: DD simulation results versus predictions of the KME model for the flow stress in specimens with: (a) $D = 0.2$ μm; (b) $D = 0.8$ μm; (c) $H = D.2$ μm; and (d) $D = 12.8$ μm.

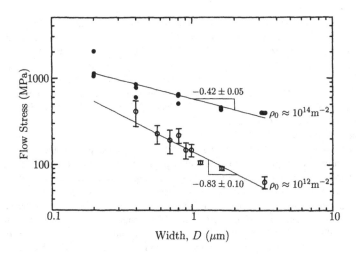

Figure 6: Scaling of the flow stress with specimen width for two values of the source density ρ_0. Error bars have height equal to one standard deviation ($\approx$ 10 calculations per datum set). Results for $\rho_0 = 1.5\,10^{14}\mathrm{m}^{-2}$ are taken from [7].

terial sample can be identified with the intrinsic constitutive response of the material only if the sample is statistically homogeneous. With respect to plastic behavior, the latter condition entails that the samples must contain a sufficient number of mobile dislocations sustaining a steady flux upon imposition of a given strain rate. Evidently, the condition of statistical homogeneity is not met in many pillar experiments as well as the low-density "2.5D" DD calculations presented here. Yet, one can make sense of experimental and simulation results in an ensemble-average sense while accepting the existence of outliers or far-from-average data points.

From the physics point of view, size effects can be associated with either the initial or evolving dislocation structures, as dictated by dislocation multiplication and intersection processes. Gradient theories of plasticity invariably invoke length scales that appear for mere dimensional consistency with no insight on the physical processes involved. For the same reason, such ad hoc length scales are generally fixed. In actuality, there is a spectrum of length scales that evolve with cumulated plastic deformation.

In principle, discrete dislocation dynamics simulations circumvent all shortcomings highlighted above. Admittedly, this is done at the cost of computational time. In addition, within the 2.5D paradigm some idealizations of actual 3D behavior are introduced, which make it impossible to capture many details regarding the dislocation structures and patterns that form during plastic deformation. However, the range of phenomena emerging from the simulations reviewed in this paper illustrate the capabilities of the 2.5D DD paradigm, as related to variations in assumed initial dislocation densities.

While the vast majority of micro- and nano-pillar experiments conducted to date corroborate the existence of strong size effects on material strength, some discrepancies remain. Among these one may list the inconsistent reporting of size effects on work hardening rates and the scaling of flow strength with pillar diameter. The present discrete dislocation dynamics analyses can elucidate some of these discrepancies and shed some light on the connection between microscopic mechanisms and overall behavior. This investigation set out to do so using the paradigm of 2.5D DD which has been shown to deliver good predictions in the bulk limit (Taylor law; athermal hardening rate as 1/100 of the elastic shear modulus; and refinement of dislocation structure.) It is within that same framework that deviations from bulk behavior have been investigated. There are at least three different origins of the plasticity size effect: (i) the samples contain enough dislocations and the size effect is due to dimensional constraints on the conventional deformation processes governed by dislocation intersections etc.; (ii) the samples contain too few dislocations to be statistically homogeneous with respect to plastic behavior: the size effect is the signature of new phenomena dictated by truncations in source length distribution, source dynamics and rare events; (iii) the samples are initially-free of dislocations: if any size effect prevails it must be associated with the nucleation, not multiplication, of dislocations from free surfaces.

The predicted range of scaling exponents in Fig. 6 encompasses the range of values measured experimentally. However, care must be taken when quantitative comparison is made between simulations and experiments. Elements worthy of consideration include: (i) single versus multislip orientation; (ii) strain level at which the scaling exponent is derived; this matters particularly in the presence of hardening; (iii) range of specimen size; (iv) the uncertainty associated with initial dislocation density and source attributes; and (v) crystal structure. For sub-micron diameter FCC crystals potentially oriented for multislip, flow-stress scaling exponents close to -1.0 are inferred from the data of [2]. The value of -0.83 predicted using 2.5D DD in a comparable size range (Fig. 6) is very close to the experimental values. However, those values are predicted only in the low initial source density case. In the high density case, the predicted scaling exponent varies with cumulated plastic strain. Few experiments, if any, have been conducted with such high density. Frick et al. [3] and Kiener et al. [20] reported a consistent effect of pillar diameter on the work-hardening rate of FCC crystals oriented for multislip. This is in qualitative agreement with the 2.5D DD predictions of Fig. 3. In addition, the scaling of the hardening rate with pillar diameter measured in [3] in the sub-micron range (slope of -1.0 in a log-log plot) is close to that predicted by [7] (slope of -0.85).

CONCLUSIONS

1. In the bulk limit, the 2.5D DD simulation paradigm delivers hardening (as opposed to strictly 2D models) with a rate of $\mu/100$; it picks up the scaling of flow stress with the square root of the dislocation density and the refinement of the dislocation structure that accompanies the increase in flow stress. Having some confidence in that model, size effects emerge under two circumstances: low and high initial dislocation density.

2. In the low density case, a regime dominated by exhaustion hardening emerges. The size effect is associated with the initial dislocation structure, the dynamic operation of isolated sources and rare events, such as junction formation.

3. In the high density case —a regime dominated by conventional forest-hardening— the size

effect is rooted in the evolving dislocation structure. Also, the flow stress scaling with specimen size is weaker, the stress–strain response is smoother and scatter is smaller.

Acknowledgments

The author acknowledges support from the National Science Foundation under grant CMMI-0748187 and from Lawrence Livermore National Security, LLC under Master Task Agreement No. B575363.

REFERENCES

[1] M. D. Uchic, D. M Dimiduk, J. N. Florando, and W. D. Nix. Sample dimensions influence strength and crystal plasticity. *Science*, 305:986–989, 2004.

[2] J. R. Greer, W. C. Oliver, and W. D. Nix. Size dependence of mechanical properties of gold at the micron scale in the absence of strain gradients. *Acta Mater.*, 53:1821–1830, 2005.

[3] C. P. Frick, B. G. Clark, S. Orso, A. S. Schneider, and E. Arzt. Size effect on strength and strain hardening of small-scale [111] nickel compression pillars. *Mater. Sci. Eng.*, 489:319–329, 2008.

[4] D Kiener, W Grosinger, G Dehm, and R Pippan. A further step towards an understanding of size-dependent crystal plasticity: In situ tension experiments of miniaturized single-crystal copper samples. *Acta Mater.*, 56:580–592, 2008.

[5] A. A. Benzerga and N. F. Shaver. Scale dependence of mechanical properties of single crystals under uniform deformation. *Scr. Mater.*, 54:1937–1941, 2006.

[6] D. S. Balint, V. S. Deshpande, A. Needleman, and E. Van der Giessen. Size effects in uniaxial deformation of single and polycrystals: a discrete dislocation plasticity analysis. *Modelling Simul. Mater. Sci. Eng.*, 14:409–422, 2006.

[7] P. J. Guruprasad and A. A. Benzerga. Size effects under homogeneous deformation of single crystals: A discrete dislocation analysis. *J. Mech. Phys. Solids*, 56:132–156, 2008.

[8] S. I. Rao, D. M. Dimiduk, T. A. Parthasarathy, M. D. Uchic, M. Tang, and C. Woodward. Athermal mechanisms of size-dependent crystal flow gleaned from three-dimensional discrete dislocation simulations. *Acta Mater.*, 56:3245–3259, 2008.

[9] J. A. El-Awady, S. B. Biner, and N. M. Ghoniem. A self-consistent boundary element, parametric dislocation dynamics formulation of plastic flow in finite volumes. *J. Mech. Phys. Solids*, 56:2019–2035, 2008.

[10] A. A. Benzerga, Y. Bréchet, A. Needleman, and E. Van der Giessen. Incorporating three-dimensional mechanisms into two-dimensional dislocation dynamics. *Modelling Simul. Mater. Sci. Eng.*, 12:159–196, 2004.

[11] A. A. Benzerga. An analysis of exhaustion hardening in micron-scale plasticity. *Int. J. Plasticity*, 24:1128–1157, 2008.

[12] A K Faradjian, L H Friedman, and D C Chrzan. Frank-Read sources within a continuum simulation. *Modelling Simul. Mater. Sci. Eng.*, 7:479–494, 1999.

[13] A. A. Benzerga. Micro-Pillar Plasticity: 2.5D Mesoscopic Simulations. *J. Mech. Phys. Solids*, 2009. Submitted.

[14] E. Van der Giessen and A. Needleman. Discrete dislocation plasticity: a simple planar model. *Modelling Simul. Mater. Sci. Eng.*, 3:689–735, 1995.

[15] D. Gómez-Garcia, B. Devincre, and L. P. Kubin. Dislocation patterns and the similitude principle: 2.5D mesoscale simulations. *Phys. Rev. Lett.*, 96:125503, 2006.

[16] B. Devincre and L. P. Kubin. Simulations of forest interactions and strain hardening in FCC crystals. *Modelling Simul. Mater. Sci. Eng.*, 2:559–570, 1994.

[17] V. V. Bulatov, F. F. Abraham, L. Kubin, B. Devincre, and S. Yip. Connecting atomistic and mesoscale simulations of crystal plasticity. *Nature*, 391:669–672, 1998.

[18] P. J. Guruprasad and A. A. Benzerga. A phenomenological model of size-dependent hardening in crystal plasticity. *Philos. Mag.*, 88:3585–3601, 2008.

[19] U. F. Kocks and H. Mecking. Physics and phenomenology of strain hardening: the FCC case. *Prog. Mater. Sci.*, 48:171–273, 2003.

[20] D Kiener, C. Motz, and G Dehm. Dislocation-induced crystal rotations in micro-compressed single crystal copper columns. *J. Mater. Sci.*, 43:2503–2506, 2008.

Mater. Res. Soc. Symp. Proc. Vol. 1185 © 2009 Materials Research Society   1185-II06-15

Measuring Elastic Properties of Highly Metastatic Cells Using Nano-Capillary Wrinkling

Nan Iyer, Katelyn Cooper, Jianing Yang, and Frederic Zenhausern
Arizona State University, The Biodesign Institute, Applied Nanobioscience, Tempe, AZ 85284

ABSTRACT

Measuring elastic properties of cells has gained importance in the study of malignant transformations. The stiffness of a cell, which is technically referred to as the modulus of elasticity or Young's Modulus, E, is the measure of the amount of cell deformation caused by an applied known force. In vitro studies have shown that cancer cells have much lower elastic stiffness than normal cells. These stiffness measurements and their differences can be used to study the behavioral mechanics of how cancer cells grow, profligate, and die in a patient. Another important use of this difference in elasticity is in cancer detection.

In this study, we explore the viability of measuring the elastic modulus of cancer cells by using a method that only requires the use of a low magnification microscope and a digital camera. In particular we are interested in applying the previously reported relationship between the wrinkling of thin films and the elastic properties of freely floating polystyrene (PS) films. Our work extends the scope of previous thin film studies by evaluating wrinkle formation in floating polystyrene films coated with biological cells. Our results show that the wrinkle formation is modified, both in morphology and in size, by the presence of a cellular monolayer on top of the PS film.

INTRODUCTION

Measuring elastic properties of cells has gained significant importance in the study of malignant transformations. *In vitro* studies have shown that cancer cells have much lower elastic stiffness than normal cells[1,2]. This finding has been further substantiated by *ex vivo* measurements that show that live metastatic cancer cells taken from the lung, chest and abdominal cavities of the patient are nearly 70% less stiff than benign, normal cells that live in the same cavities[3]. These stiffness measurements and their differences can be used to study the behavioral mechanics of how cancer cells grow, profligate, and die in a patient. Another important use of this difference in the elasticity of cancerous and normal cells is in cancer detection. Measuring elasticity of cells could be used as a mechanical marker for detecting cancer in the clinic and to complement current methods such as microscope examination of tissue samples and antibody labeling which are not always accurate due to the visual similarity between sick and normal cells. For example, there is a critical clinical need for whether an early skin lesion is a melanoma, and if so, the likelihood of the lesion to metastasize. Any change in cell elasticity may be a key component for malignant melanoma to acquire its full metastatic potential.

The stiffness of a cell, which is technically referred to as the modulus of elasticity or Young's Modulus, E, is the measure of the amount of cell deformation caused by an applied known force. In the above mentioned *in vitro* and *ex vivo* studies, an AFM (Atomic Force Microscope) was used to measure the modulus of elasticity of cells. A sharp probe located at the end of the AFM cantilever was used to probe and deform the cell with a known applied normal force. The deflection of the cantilever, which is a reflection of the amount of deformation in cell, is measured, and the elastic modulus is then calculated using the applied force value. However, this method is expensive, complex and does not lend itself to expeditious diagnosis as required in the clinic. Furthermore, AFM imaging is limited to very small scan areas and single cell analysis. This may also prevent access to cellular phenotypic and physical events that may occur in ECM interaction at a larger scale.

This work is a preliminary study that explores the viability of measuring the elastic modulus of cancer cells by using a simple yet powerful method that only requires the use of a low magnification microscope and a digital camera. The proposed method utilizes the phenomenon of wrinkling of thin films such as can be seen on our skin as it is stretched by smiling or the skin of fruit as it dries. In particular we are interested in extending a study that reported the use of wrinkling of thin films to measure the elastic properties of freely floating polystyrene (PS) films[4]. We extend this theory by measuring wrinkle formation in floating polystyrene films coated with biological cells. The elastic modulus, E, of the cells may then be determined by manipulating the appropriate equations.

THEORY

A floating thin film will often deform out of plane to form wrinkles when subject to a normal loading force. Application of a drop of water on the surface of spin coated polystyrene films, tens of nanometers in thickness, induced a wrinkling pattern on the film due to capillary forces. A semi-empirical scaling relationship was obtained from considering the energy minimization of the bending transverse to the wrinkles and the stretching along their length[4]. Therefore, measuring the length, L, and number of wrinkles, N, allows the determination of the Young's modulus, E, and the thickness, h, of the floating film. The equations of interest are given by

$$N = C_N \left[\frac{12\left(1 - \Lambda^2\right)}{E} \right]^{\frac{1}{4}} a^{\frac{1}{2}} h^{-\frac{3}{4}} \tag{1}$$

$$L = C_L \left(\frac{E}{\gamma} \right)^{\frac{1}{2}} a h^{\frac{1}{2}} \tag{2}$$

where Λ is the Poisson ratio, γ is the surface energy, and a is the radius of the drop causing normal loading. C_N and C_L are numerical constants determined from plotting the relevant experimental data and using a curve-fit.

We extend this theory by measuring wrinkle formation in floating polystyrene films coated with biological cells. By counting the number of wrinkles, N, and the length of wrinkles, L, the elasticity of thin PS films with cell coatings may be determined. Manipulation of the appropriate equations will then allow for determination of the elastic modulus, E, of the cells.

EXPERIMENT

Figure 1 shows the experimental set-up that is being used for this study. The experimental procedure is modeled after the standard protocol established previously[4]. Figure 2 illustrates these steps. Briefly, a thin layer of polystyrene is spin coated onto a glass slide. A circle is scribed on the PS film and when the slide is submerged in water, the scribed PS circle lifts releases from the slide and floats to the surface of the water. A droplet of water is pipetted onto the floating PS film and wrinkles are formed.

For observation with cells, the PS coated glass slides are rinsed with a quick stream of de-ionized water followed by 70% ethanol. The slides are air dried in the culture hood before being subjected to UV light sterilization for approximately 2 hours. A monolayer of cervical cancer cells is then cultured on the PS film and the slide is placed in an incubator until the cells exhibit approximately 70-80% confluence as determined by optical microscopic inspection. The cultured slide is then scribed and submerged in water. In the case of the cellular monolayer, the floating procedure was not as consistent as the case of the untreated PS film, and thus a greater number of trials are necessary to achieve the same yield of floating thin films.

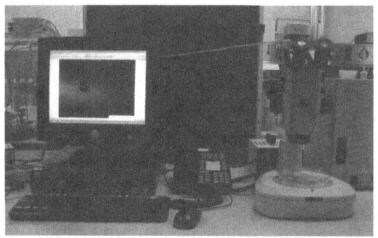

Figure 1: Low magnification Nikon microscope with diascopic base and digital camera used to capture wrinkling phenomenon.

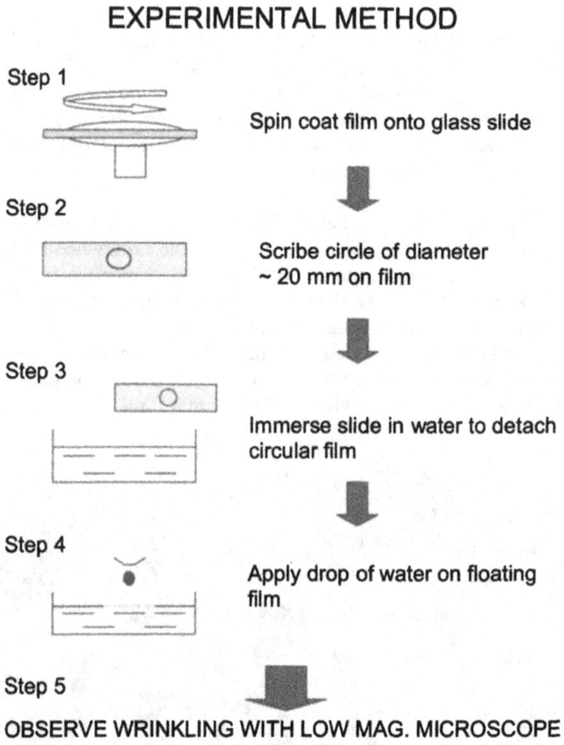

EXPERIMENTAL METHOD

Step 1

Spin coat film onto glass slide

Step 2

Scribe circle of diameter
~ 20 mm on film

Step 3

Immerse slide in water to detach
circular film

Step 4

Apply drop of water on floating
film

Step 5

OBSERVE WRINKLING WITH LOW MAG. MICROSCOPE

Figure 2: Experimental protocol used to create floating thin films with observable capillary wrinkling

RESULTS

Experiments were conducted on a PS thin film (93nm), with and without a cell culture on top. In each case, wrinkling was observed consistently. Figure 3 below is a photograph taken of a typical wrinkling pattern observed on a floating PS film due to the weight of a drop of water. The drop of water, 1 mm in diameter, was pipetted onto the surface of the film. The wrinkles generated can be clearly seen and the number of wrinkles, N, and the length of wrinkles, L, may be measured. In this case for a PS film of thickness, h = 93 nm, N = 45 and L = 1.3 mm. Substituting these values into equation 2, we obtain E = 3.14 ± 0.18 GPa for the elastic modulus of the PS thin film. The values for C_L = 0.031 and γ = 72 ± 0.3 mN/m were obtained from literature[4] and a = 0.5 mm, is the radius of the drop of water. This calculated value of E is within the range (~3.3 GPa) of the accepted elastic modulus value for PS[5].

Figure 4 shows the analogous wrinkling pattern observed on the PS film material, but in this case with a layer of cells cultured on the surface. Clearly, there is some change in the wrinkling pattern due to the presence of the cells on the PS surface. The symmetric pattern of wrinkles is not as uniform, and the number and size of wrinkles has changed as well.

Figure 3: Wrinkling pattern on PS film thickness 93 nm loaded with a 1 mm diameter drop of water

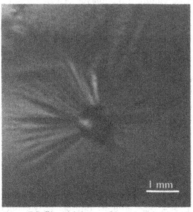

Figure 4: Wrinkling pattern on PS film thickness 93 nm with monolayer of cells loaded with a 1 mm diameter drop of water.

In the cell loaded case, the wrinkles are generally larger and fewer, though the distribution of wrinkles sizes seems to vary from 1 mm to 3 mm. Due to the non-uniformity of the wrinkling pattern, it is difficult to calculate N and L. However, the wavelength of the wrinkles, λ, was measured at a distance, $d = 0.85$ mm from the center of the bead of water. Using equation 3, the number of wrinkles, N, is estimated and compared for Figures 3 and 4. The results are outlined in Table 1 below.

$$N = 2\pi d / \lambda \tag{3}$$

Table 1: Comparison between pure PS film and cell coated PS film

	λ @ 0.85 mm radial distance	N (measured)	N (calculated using Equation 3)	L (measured)
PS film (Figure 3)	0.11 ± 0.01 mm	45	53	1.3 mm
Cell coated PS film (Figure 4)	0.17 ± 0.01 mm	-	31	Range from 1-3 mm

In general, the size of normal and cancerous cervical cells, depending on where the cell is harvested from in the cervix, ranges from 100-500 times[6] the thickness of the PS film. However, the elastic modulus of PS film is approximately 10,000 times[3] higher than such cells. These competing effects result in a relatively modest change in the wrinkling mechanics.

CONCLUSIONS

An experimental procedure to culture a monolayer of cells onto a polystyrene (PS) thin film, float the film in a bath of water, and to conduct capillary wrinkling tests on the film, was developed. The results to date confirm that the presence of a cellular monolayer modifies the wrinkling pattern of the PS film, both in terms of morphology of the wrinkling pattern and the size of the wrinkles.

Further work using this method and refinement of the experimental method using cell cultured PS films, which is currently in progress, should yield quantifiable values for cell elasticity. In addition the effect of the cell culturing protocol such as sample sterilization and culture media modification of the PS film layer on the material structure is being studied.

REFERENCES

1. Guck, J. et al. *Biophys. J.* **88**, 3689-3698 (2005)
2. Suresh, S. *Acta Biomater.* **3**, 413-438 (2007)
3. Cross, S.E., Jin, Y.-S., Rao, J.Y., Gimewski, J.K., *Nature Nanotech.*, **19(38)** (2008)
4. Huang, J., Juskiewicz, M., Jeu, W., Cerda, E., Emrick, T., Menon, N., Russell, T., *Science* **317**, 650 (2007)
5. Brandrup, J., Immergut, E.H, Polymer Handbook (Wiley, New York, ed. 3, 1987)
6. Foden, A.P., Freedman, R.S., Abrahams, C., *S. Afr. Med. J.*, **50**, 571 (1976)

Mater. Res. Soc. Symp. Proc. Vol. 1185 © 2009 Materials Research Society   1185-II07-03

Universal Scaled Strength Behaviour for Micropillars and Nanoporous Materials

Brian Derby and Rui Dou
School of Materials, University of Manchester,
Grosvenor Street, Manchester, M1 7HS, UK

ABSTRACT

The strength of submicron fcc structure metal columns, σ, fabricated by FIB machining or electrodeposition, shows a strong correlation with specimen diameter, d, with $\sigma/\mu = A(d/b)^{-0.63}$, where A is a constant, μ is the single crystal shear modulus resolved onto the slip system and b is the Burgers' vector. The strength of bcc structure metals does not follow such a well defined correlation with size across different metals but the data occupies the same region of parameter space as with the fcc metals. Nanoporous gold specimens show a similar size-correlated behaviour but with an exponent of -0.5. This may indicate different mechanisms operating in each case.

INTRODUCTION

The strength of metal pillars, columns or wires with diameter in the range of 30 nm – 30 μm shows a pronounced size effect, with yield strength in excess of 1 GPa for the smallest pillars tested [1-5]. A similar behavior is seen with the deformation of nanoporous metals, with similar strength values reported when the ligament diameter approaches 10 nm [6-10]. Previous work has investigated this behavior and proposed a simple scaling relation for the behaviour of face centred cubic (fcc) structure metals [11,12]. Here we explore this scaling relation and extend it to materials with other crystal structures. We also investigate whether nanoporous gold scales in the same manner.

A large number of experiments have been reported on micropillar compression studies of a range of metals including Au, Ni, Cu, Al, Mo and the semiconductor GaAs. [1-5, 13-21] In most cases the specimens were small single crystal specimens with aspect ratios (height/diameter) ≈ 3. In the majority of the reports, the deformation of individual micropillars follows an erratic stress/strain history with deformation occurring in bursts of strain at almost constant stress interspersed by regions of elastic deformation. In these samples deformation appeared to be highly localized with clearly visible shear offsets along the deformed pillar. In a few cases deformation showed a continuous increase in deformation stress with increasing plastic strain, analogous to the deformation behavior of polycrystals and in these cases deformation was not localized..

At small length scales conventional mechanisms for the generation of dislocations are severely constrained by the small dimensions of the specimens, which either physically limit the length of dislocation segments available for dislocation multiplication or remove mobile dislocations through the close proximity of free surfaces and consequent attractive mirror forces. [14] This "dislocation starvation" strengthening mechanism is supported by experimental observations on the behavior tested in compression, where deformation is observed occurring in bursts and a deformed column shows extensive stepping on the column surface. Transmission electron microscopy (TEM) observations have found very low dislocation densities in gold columns with diameter < 100 nm deformed to macroscopic strains in excess of 30% [5]. Shan et al has observed a decrease in dislocation density during compression deformation of a Ni nanopillar observed *in situ* in a TEM. [22] Frick et al however, [16] found significant dislocation densities in deformed Ni micropillars. Molecular dynamics simulations predict the nucleation of dislocations during deformation. These model some of the features of nanopillar deformation, such as deformation in slip bursts, but do not give adequate quantitative predictions. [23,24].

SCALING EXPERIMENTAL DATA

If we assume that the deformation of micropillars is controlled by dislocation motion in its broadest sense. It is reasonable, therefore, to expect that any strengthening is caused by some interaction of the dislocation with some feature of the micropillar. In which case, the controlling physical property of the dislocation will be its self energy or line tension. The self energy of a dislocation is determined by its Burgers' vector and the elastic modulus of the material. In figure 1 we present the data from a number of literature sources for the compression flow stress of fcc micropillars in a form where σ_{crss} is normalized by the material shear modulus resolved onto the appropriate crystallographic slip system for mobile dislocations, μ, and the pillar diameter, d, is normalized by the Burgers vector of the dominant slip system, b. This data shows a very strong empirical correlation between the normalized pillar diameter and the resolved flow stress.

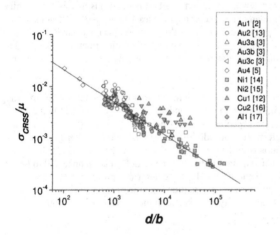

Figure 1 Critical resolved shear stress for FCC micropillar deformation, normalized by shear modulus, plotted against pillar diameter, normalized by Burgers' vector, for a number of fcc metals.

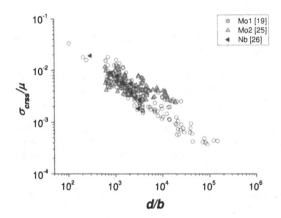

Figure 2 Critical resolved shear stress for bcc micropillar deformation, normalized by shear modulus, plotted against pillar diameter, normalized by Burgers vector, for a number of bcc materials (filled symbols). The fcc data from figure 1 are displayed as open symbols.

Recently there have been a small number of studies published for the micropillar compression of body centred cubic (bcc) structure metals. This data appears to show less consistency within the material class than has been observed with fcc metals. As far as we are aware, there has been no consideration of the scale dependence of strength in bcc materials in its normalized form. In figure 2 we present such a normalization for strength data from FIB machined specimens of Mo and Nb with diameter < 1000 nm [19,25,26]. The data for fcc metals is plotted on the same diagram (open symbols) for comparison purposes. It is clear that the normalized data occupies a similar region of the parameter space as is seen with fcc metals. However, the exponent of the scaled relation for Mo and Nb (gradient in the log-log plot) are different, and both are different from the mean exponent of the fcc data.

DISCUSSION

The data presented in normalized form in figure 1 shows a consistent trend of strength as a function of pillar diameter for fcc metal specimens produced by FIB machining and electroplating into templates; with the data for Au, Ni and Al occupying the same region of parameter space. This is remarkable considering the likely errors and uncertainties in each of the individual sets of experimental data that are plotted. The critical resolved shear stress for plastic flow can be related to the pillar diameter, material shear modulus and the Burgers' vector of the active slip system with the following empirical relation:

$$\frac{\sigma_{crss}}{\mu} = A\left(\frac{d}{b}\right)^{m} \qquad (1)$$

The solid line in figure 1 shows the line obtained by a regression fit to the data obtained from Al, Au and Ni micropillars, this has an exponent of $m = -0.63$.

Figure 2 shows the size-scaled behavior for Mo and Nb specimens normalized by shear modulus and Burgers' vector. The gradient of the Mo and Nb data sets are different both from each other and from the fcc data with Nb $\approx$ -1.0 [26] and Mo $\approx$ -0.33 [19,25]. However, the normalized strength and pillar diameter of the bcc metals is clearly superimposed on the range of data reported for fcc materials (Figure 2).. This similarity in strength behavior is remarkable given that bcc metals are believed to show different deformation mechanisms from fcc metals in terms of dissociated dislocation structures, which in Mo gives different mobility for edge and screw dislocations.

Finally we return to the behavior of nanoporous gold. In an earlier paper [5] we compared the reported strengths of nanoporous gold and gold micropillars tested in compression without resolving the stresses onto slip planes. Nanoporous gold is a polycrystalline material with each ligament in the structure being an individual grain. If we assume that these ligaments are oriented randomly in space, then we can use the Sachs factor (the average Schmid factor for a random assembly of fcc grains) as a suitable constant to determine the mean resolved shear stress in plastically deforming nanoporous gold. This resolved shear data is normalized by Burgers' vector and shear modulus and presented in figure 3 along with the data for gold micropillar compression.

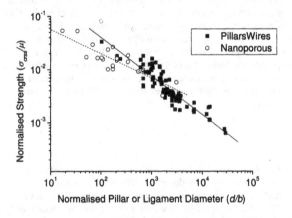

Figure 3 Critical resolved shear stress for gold micropillar/nanowire deformation (solid symbols) and nanoporous gold (open symbols), normalized by shear modulus, plotted against pillar diameter, normalized by Burgers vector.

From figure 3 it is clear that the resolved plastic shear stress for gold micropillar compression and for the deformation of nanoporous gold occupy the same region of parameter space. The solid line shows the best straight line through the gold pillar data, which has an exponent of -0.6, consistent with figure 1, the dashed line shows the best fit through the nanoporous gold data and this has an exponent close to -0.5.

A possible explanation for the different exponents found for the size-scaled behavior of the nanoporous gold lies in the different deformation processes that occur in each material. The deformation of micropillars is believed to occur through the nucleation and escape of dislocations leading to localized shear banding. However, nanoporous gold is believed to deform by a plastic collapse mechanism proposed by Gibson and Ashby [27]. This mechanism requires the action of plastic hinges and these contain large gradients of plastic strain that lead to a well known scale dependence of strength at small length scales with an exponent of -0.5 in bend deformation [28,29].

CONCLUSIONS

The strength data for micropillar compression of FIB produced FCC metal specimens are seen to lie on a common trend line with a power dependence of approximately -0.6, when the data is resolved onto the highest Schmid factor slip system and it is normalized by shear modulus and Burgers' vector. This correlation may possibly extend to materials with crystal structures other than the FCC metals. The behavior of nanoporous metals is similar to that of micropillars but they may show a different exponent that describes their size dependent behavior. This difference between the behavior of the micropillars and the nanoporous gold may indicate a different deformation mechanism in these two materials.

REFERENCES

1. Uchic, M.D., Dimiduk. D.M., Florando, J.N. & Nix, W.D., *Science* **305**, 986-989 (2004).
2. Greer, J.R., Oliver, W.C. & Nix, W.D., *Acta Mater.* **53**, 1821–1830 (2005).
3. Volkert, C.A. & Lilleodden, E.T., *Philos. Mag.* **86**, 5567-5579 (2006).
4. Maass, R., Grolimund. D., Van Petegem. S., Willimann, M., Jensen, M., Van Swygenhoven, H., Lehnert, T., Gijs, M.A.M., Volkert, C.A., Lilleodden, E.T. & Schwaiger, R., *Appl. Phys. Lett.* **89**, 151905 (2006).
5. Dou, R. & Derby, B., *Scripta Mater.* **59**, 151-154 (2008).
6. Biener, J., Hodge, A.M., Hamza, A.V., Hsiung, L.M. & Satcher, J.H., Nanoporous *J. Appl. Phys.* **97**, 024301 (20050.
7. Volkert, C.A., Lilleodden, E.T., Kramer, D. & Weissmuller, J., *Appl. Phys. Lett.* **89**, 061920 (2006).
8. Lee, D., Wei, X., Chen, X., Zhao, M., Jun, S.C., Hone, J., Herbert, E.G., Oliver, W.C. & Kysar, J.W., *Scripta Mater.* **56**, 437-440 (2007).
9. Hakamada, M., & Mabuchi, M., *Scripta Mater.* **56**, 1003-1006 (2007).

10..Hodge, A.M., Biener, J., Hayes, J.R., Bythrow, P.M., Volkert, C.A. & Hamza, A.V., *Acta Mater.* **55**, 1343-1349 (2007).
11. Uchic M.D., Shade, P.A. and Dimiduk, D.M. *JOM*, **61(3)**, 36 (2009).
12. Dou, R., & Derby, B., *Scripta Mater.* **61**, 524-527 (2009).
13. Kiener, D., Motz, C., Schöberl, T. Jenko, M. & Dehm, G., *Adv. Eng. Mater.***8**, 1119-1125 (2006).
14. Greer, J.R. & Nix, W. D., *Phys. Rev. B* **73**, 245410 (2006).
15. Dimiduk, D.M., Uchic, M.D. & Parthasarathy, T.A. *Acta Mater.* **53**, 4065–4077 (2005).
16. C.P. Frick, B.G. Clark, S. Orso, A.S. Schneider, E. Arzt, *Mater. Sci. Eng. A* **489**, 319–329 (2008).
17. Kiener, D., Motz. C, & Dehm, G., et al, *J. Mater. Sci.* **43**, 2503–2506, (2008).
18. Ng, K.S. & Ngan, A.H.W *Acta Mater.* **56**, 1712–1720 (2008).
19. Kim, J.Y., and Greer, J.R., *Appl. Phys. Lett.* **93**, 101916 (2008).
20. Bei, H., Shim, S., George, E.P., Miller, M.K., Herbert, E.G. & Pharr, G.M., *Scripta Mater.* **57**, 397-400 (2007).
21. J. Michler, K. Wasmer, S. Meier, and F. Östlunda, *Appl. Phys. Lett.* **90**, 043123 (2007).
22. Shan, Z.W., Mishra, R.K., Asif, S.A.S., Warren, O.L. & Minor A.M., *Nature Mater* **7**, 115-119 (2008).
23. Rabkin E., Nam H.S. & Srolovitz D.J., *Acta Mater.* **55**, 2085-2099 (2007).
24. Rabkin, E & Srolovitz D.J., *Nano Lett.* **7**, 101-107 (2007).
25. Schneider, A.S., Clark, B.G., Frick, C.P., Gruber, P.A. & Arzt, E., *Mater. Sci. Eng. A* **508**, 241-246 (2009).
26. Kim, J.-Y., Jang, D., & Greer, J.R., *Scripta Mater.* **61**, 300-303 (2009).
27. Gibson, L.J., & Ashby, M.F., *Proc. Royal Soc. Lon. A* **382A**, 43-59 (1982).
28. Fleck, N.A., Muller, G.M., Ashby, M.F., Hutchinson, J.W., *Acta Metall. Mater.* **42**, 475-487 (1994).
29. J.S. Stolken, A.G. Evans, *Acta Mater.* **46**, 5109-5115 (1998).

Mater. Res. Soc. Symp. Proc. Vol. 1185 © 2009 Materials Research Society 1185-II07-04

Correlation Between Activation Volume and Pillar Diameter for Mo and Nb BCC Pillars

A. S. Schneider[1,Ω], B. G. Clark[2], C. P. Frick[3], and E. Arzt[4]

[1]Max Planck Institute for Metals Research, Heisenbergstrasse 3, 70569 Stuttgart, Germany
[2]Sandia National Laboratories, Albuquerque, NM 87185, USA
[3]University of Wyoming, Mechanical Engineering Department, 1000 East University Avenue, Laramie, WY 82071, USA
[4]INM - Leibniz Institute for New Materials and Saarland University, Campus Building D2 2, 66123 Saarbrücken, Germany

ABSTRACT

Compression tests with varying loading rates were performed on [001] and [235] oriented small-scale bcc Mo and Nb pillars to determine the contribution of thermally activated screw dislocation motion during deformation. Calculated activation volumes were shown to be in the range of $1.3 - 8.8\ b^3$ and by further examination were found to decrease with pillar diameter. This suggests that the kink-pair nucleation of screw dislocations is enhanced by surface effects in the micron and submicron range.

INTRODUCTION

Compression tests on focused ion beam (FIB) machined, single crystal metal micropillars have shown that flow stress is inversely related to pillar diameter [1-6]. Although the fundamental mechanisms which govern the size effect are still under debate, a consistent scaling relationship for the flow stress σ_y with pillar diameter d on the order of $\sigma_y \propto d^{-0.6} - d^{-1.0}$ has been found for several face-centered cubic metals [1-5]. Because fcc and bcc metals differ fundamentally in their dislocation processes it is perhaps not surprising that bcc metals show a different size scaling on the order of $\sigma_y \propto d^{-0.2} - d^{-0.4}$ [7-8]. The weaker size dependence of bcc metals has been attributed to the low mobility of screw dislocations in bcc metals leading to either conventional dislocation-dislocation interactions [7] or kinetic pile-ups of screw dislocations in the vicinity of dislocation sources [8]. To elaborate on the role of thermally activated screw dislocation motion in the deformation of small-scale bcc metal pillars, compression tests on Mo pillars were performed at various loading rates in a previous study [8]. Calculated activation volumes were found to be in the range of $1.3 - 5.3\ b^3$, where b is the Burgers vector, which is in good agreement with activation volumes measured for thermally activated kink-pair nucleation of screw dislocations in bulk bcc Mo [9]. In this study the relationship between pillar diameter and calculated activation volumes are examined in more detail for Mo, and compared to new results from Nb pillars.

EXPERIMENTAL

For this study, [001] and [235] oriented Mo and Nb pillars with diameters between 300 nm and 3000 nm were FIB machined on the surfaces of bulk Mo and Nb single crystals using a Dual Beam™ FIB. Both orientations for each material were obtained from the same bulk single crystal by electron discharge machining after the orientations were first determined by Laue-diffraction. For each orientation several diameters were investigated and for each diameter about 24 pillars were prepared. The pillars were tested with loading rates between 1 and 500 μN/s. For consistency, top pillar diameters were used to calculate the engineering stress-strain and activation volumes. Further experimental details can be found in [8].

DISCUSSION

The results of the loading rate tests for the [001] and the [235] oriented Mo and Nb pillars are shown in Figure 1a and Figure 1b, respectively. Both materials show an increase in stress with increasing loading rate independent of pillar diameter. This time-dependent deformation behavior is known to correlate with the presence of thermally activated deformation processes.

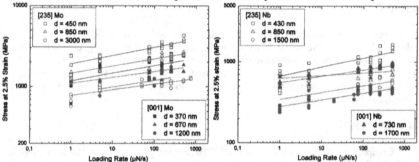

Figure 1. (a) Loading rate dependence for [001] and [235] oriented Mo pillars with diameters ranging from 370 nm to 3000 nm. (b) Loading rate dependence for [001] and [235] oriented Nb pillars with diameters ranging from 430 nm to 1500 nm. Time-dependent deformation behavior indicates the presence of thermally activated deformation processes.

Activation volume calculation

To calculate the activation volumes based on the results shown in Figure 1, the loading rate was first converted to an elastic strain rate by dividing it by cross-sectional area and elastic modulus. For these inherently load controlled experiments the elastic strain rate was used as an approximation for the real strain rate. This seems to be valid since the stresses are measured at very low strains, and it can be assumed that the deformation up to this point is nearly elastic. The activation volume v was calculated based on the equation [10]:

$$v = mkT\left(\frac{\partial \ln \dot{\varepsilon}}{\partial \sigma}\right)$$

(1)

where m is the Taylor factor (2.128 for the [001] and 2.222 for the [235] orientation), k is Boltzmann's constant, T is temperature, $\dot{\varepsilon}$ is strain rate, and σ is stress.

According to Equation 1, the slope obtained from plotting the natural log of the strain rate versus stress, as demonstrated in Figure 2 for the [235] oriented Mo pillars, is proportional to the activation volume. For Mo pillars, calculated activation volumes were found to be in the range of $1.3 - 5.3\ b^3$, while for Nb pillars slightly higher values were obtained ($2.2 - 8.8\ b^3$). The calculated activation volumes fall in the range of 1.3 – 8.8 b^3 indicating that, as in bulk bcc metals, the kink nucleation of screw dislocations is the rate controlling deformation mechanism in bcc pillars. As the rate controlling deformation mechanism is the same for both materials and all tests were performed at room temperature both materials show very similar values for the calculated activation volumes.

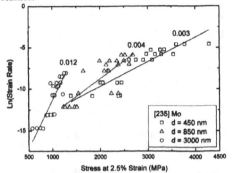

Figure 2. Natural log of elastic strain rate versus stress at 2.5% strain for [235] oriented Mo pillars. The slope of each curve is proportional to the activation volume for that pillar diameter (see Equation 1).

Correlation of activation volume with pillar diameter

Examination of the results revealed that the calculated activation volumes are strongly correlated to the pillar diameter, as can be seen in Figure 3. The Mo pillars as well as the Nb pillars show decreasing activation volumes with decreasing pillar size. This indicates that with decreasing pillar size less thermal activation is required to nucleate kinks. It is likely that for small dimensions surface effects enhance kink nucleation, as recently proposed by Weinberger and Cai [11]. In their molecular dynamics (MD) simulations they have shown that in bcc Mo image forces can lead to the nucleation of kinks at the pillar surface. Although the sample size used for their simulations ($\leq$ 50 nm) is small compared to our pillars, it is likely that with increasing surface to volume ratio the contribution of surface effects (i.e. the image force assisted kink nucleation) increases, which could explain the decrease in activation volumes with decreasing pillar diameter.

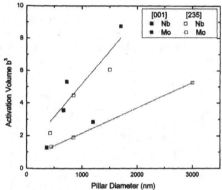

Figure 3. Activation volume versus pillar diameter for [001] and [235] oriented Mo and Nb pillars, showing that smaller activation volumes correlate to smaller pillar diameters.

Although surface effects likely contribute to the effect demonstrated in Figure 3, it is important to note that activation volume is also known to decrease with increasing stress [9]. Since stresses are higher for small pillars (due to the size effect), the decrease in activation volumes could be related to their higher stresses and may not only be caused by surface effects. Assuming this to be true, the consistently higher activation volumes of Nb compared to Mo could then be explained by the lower stresses of the Nb pillars [12]. In addition, Nb pillars show a stronger size effect in comparison to Mo pillars [12], which may be responsible for the stronger correlation of activation volume and pillar diameter for Nb. Further investigations on a broader range of diameters are required to distinguish between both contributions.

Comparison with fcc pillars

In previously published work comparing Mo pillars to fcc pillars [8], normalized stresses were found to be very similar at pillar diameters close to 200 nm. Enhanced mobility of screw dislocations and ease of kink-pair nucleation due to surface effects could explain this similarity. The difference in mobility between bcc edge and screw dislocations, which is responsible for the difference mechanical behaviors of bcc and fcc bulk metals, would be reduced due to surface assisted kink nucleation. Thus in smaller pillars, similar dislocation processes may lead to comparable strengths for fcc and bcc. For larger bcc pillars, with smaller surface to volume ratio and therefore less surface assisted kink nucleation, the difference in mobilities of screw and edge dislocations would result in a deviation in strength between fcc and bcc pillars, as observed for Mo pillars [8].

CONCLUSIONS

The decrease in activation volume with decreasing pillar diameter for bcc Mo and Nb shows that in the submicron range surface effects may enhance the mobility of screw dislocations. This observation helps explain why fcc and bcc pillars close to 200 nm have similar normalized stresses [8]. However, the influence of stress on activation volume (decrease in activation volume with increasing stress) makes it difficult to quantify this effect due to increased stresses in small pillars due to the size effect. Nevertheless, the enhanced mobility of screw dislocations for small bcc pillars is reasonable to explain the difference in size dependence and the convergence of the normalized stresses at small pillar diameters for fcc and bcc pillars.

REFERENCES

1. M.D. Uchic, D.M. Dimiduk, J.N. Florando, and W.D. Nix, Science **305**, 986 (2004).
2. J.R. Greer, W.C. Oliver, and W.D. Nix, Acta Materialia **53**, 1821 (2005).
3. D.M. Dimiduk, M.D. Uchic, and T.A. Parthasarathy, Acta Materialia **53**, 4065 (2005).
4. C.A. Volkert and E.T. Lilleodden, Philosophical Magazine **86**, 5567 (2006).
5. C.P. Frick, B.G. Clark, S. Orso, A.S. Schneider, and E. Arzt, Materials Science and Engineering A **489**, 319 (2008).
6. D. Kiener, C. Motz, T. Schoberl, M. Jenko, and G. Dehm, Advanced Engineering Materials **8**, 1119 (2006).
7. S. Brinckmann, J.-Y. Kim, and J.R. Greer, Physical Review Letters **100**, 15502 (2008).
8. A.S. Schneider, B.G. Clark, C.P. Frick, P.A. Gruber, and E. Arzt, Materials Science and Engineering A **508**, 241 (2009).
9. M. Tang, L.P. Kubin, and G.R. Canova, Acta Materialia **46**, 3221 (1998).
10. R.J. Asaro and S. Suresh, Acta Materialia **53**, 3369 (2005).
11. C. R. Weinberger and W. Cai, Proceedings of the National Academy of Sciences **105**, 14304 (2008).
12. A. S. Schneider, D. Kaufmann, B.G. Clark, C.P. Frick, P.A. Gruber, R. Mönig, O. Kraft, and E. Arzt (in preparation).

Mater. Res. Soc. Symp. Proc. Vol. 1185 © 2009 Materials Research Society   1185-II07-08

Nano-meter scale plasticity in KBr studied by nanoindenter and force microscopy

P. Manimunda[1], T. Filleter[2], P. Egberts[2,3], V. Jayaram[1], S.K. Biswas[1], and R. Bennewitz[2,3]

[1]Department of Mechanical Engineering, Indian Institute of Science, Bangalore, India
[2]Department of Physics, McGill University, Montreal, Canada
[3]INM – Leibniz Institute for New Materials, Saarbrücken, Germany

ABSTRACT

The early stages of plasticity in KBr single crystals have been studied by means of nano-meter-scale indentation in complementary experiments using both a nanoindenter and an atomic force microscope. Nanoindentation experiments precisely correlate indentation depth and forces, while force microscopy provides high-resolution force measurements and images of the surface revealing dislocation activity. The two methods provide very similar results for the onset of plasticity in KBr. Upon loading we observe yield of the surface in atomic layer units which we attribute to the nucleation of single dislocations. Unloading is accompanied by plastic recovery as evident from a non-linear force distance unloading curve and delayed discrete plasticity events.

INTRODUCTION

A prominent element of discrete plasticity is the pop-in event observed in nanoindentation experiments on many materials. It comprises the sudden creation of a multitude of dislocations and marks the transition from elastic to elasto-plastic response of the surface. Despite the singularity of the pop-in event, the resulting defect structure in the plastic zone can still be too complex for a full microscopic understanding. Various experimental strategies have been implemented to arrive at a simpler plastic response that allows for a detailed analysis. Woirgard and collaborators proceeded in their studies of MgO from a Berkovich indenter to a large-radius spherical indenter. The indentation experiments then resulted in a well-defined set of dislocation half-loops which could be quantitatively analyzed using a combination of etching techniques and atomic force microscopy imaging [1]. Gerberich and co-workers have followed nanometer-scale plasticity in great detail by studying isolated silicon nanoparticles and ultrathin metal films [2,3]. They found that few dislocations trapped in nanoparticles between the indenter and a rigid substrates cause a significant work hardening, and that the interactions between dislocations separated by only a few nanometers can drive reverse plasticity after unloading.

Our approach to enter this regime attempts to limit the number of dislocations by employing very sharp tips and very low loads in indentation experiments on single crystal samples. This strategy allows the observation of single-dislocation pop-in events. In this paper we directly compare the two experimental methods, nanoindentation and atomic force microscopy (AFM) on the example of nanometer-scale indentation into KBr single crystals.

EXPERIMENTAL METHODS

Nanoindentation studies the local elastic and plastic response of a surface to a point-like load when a sharp tip is brought into contact with the surface. Its displacement into the surface is measured as a function of time while the load is increased, held at a maximum value, and decreased again [4]. The essential parts of a nanoindentation experiment are the load control and the precise measurement of the displacement. Atomic force microscopy has also been used in nanoindentation experiments [5-8]. The tip of the AFM, which is attached to a micro-fabricated cantilever with a force constant of the order of 1-50 N/m, is brought into contact with the surface. The load is then increased by approaching the sample towards the base of the cantilever and thereby bending the cantilever. Measured force and sample approach are interdependent and, consequently, no direct measurement of the displacement of the tip into the surface is possible. This is a significant disadvantage compared to instrumented nanoindentation experiments. On the other hand, AFM-based nanoindentation experiments provide force measurements with a lower noise level and the ability to image the indented area with nanometer-scale resolution. Results of complementary experiments using nanoindenter and AFM are to be presented in different ways reflecting the differences in the two techniques. Figs. 1(a) shows a force vs. sample approach curve for an AFM-based indentation experiment. The curve is dominated by the linear response of the cantilever. Sudden jumps in the force indicate plastic events which will be discussed below. Figure 1(b) shows the result of a nanoindentation experiment. Here the force is plotted as a function of the displacement of the tip into the surface. The curve shape allows the analysis of elastic response and hardness [9], and the complete cycle of loading, holding, and unloading opens an opportunity to quantify permanent deformation of the indented surface.

In this study, we are interested in the onset of plasticity in single-crystal samples. We study indentation depths that are smaller than the apex radius of our indenting tips. Therefore, the shape of the body of the indenting tip does not affect the stress distribution in the indentation experiments. KBr(100) surfaces were prepared by cleaving the crystal along (100) plane in air using a fresh razor blade. The cleavage produces atomically flat terraces of several hundred nanometers width. Reproducible indentation results were obtained on samples cleaved within few hours before indentation experiments. Samples studied in vacuum were introduced to the vacuum chamber within minutes after cleavage in order to minimize contamination.

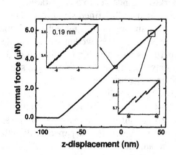 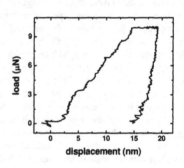

Figure 1: (a) Force vs. sample approach curve for an AFM-based indentation experiment in vacuum using a diamond tip with an estimated radius of 30 nm. Pop-in events are observed at 3.45 and 5.8 μN as detailed in the inserts. (b) Load vs. displacement curve of a nanoindentation experiment in air using a sharp diamond tip of similar tip radius. Pop-in events are observed at 3.4, 4.4, 5.2, and 6.8 μN.

RESULTS AND DISCUSSION

Plastic events during loading and unloading

In this section we will describe the results of indentations into atomically flat surfaces of KBr single crystals by means of AFM and nanoindentation. The AFM loading curve in Fig. 1(a) exhibits several sudden displacements of the tip into the sample. While the elastic displacement of the surface is difficult to quantify for reasons of non-linearity in the force sensor and the piezo actuators, the height of these sudden displacements can be determined accurately. For the sharp tips and low loads used in this study, most displacements have a height around 0.3 nm. This height corresponds to a single layer of KBr, and we conclude that each jump indicates the glide of KBr along a glide plane by one atomic distance under the stress of the indenting tip. A corroborating analysis of many such jumps has been given elsewhere [7].

The results of a comparable nanoindentation experiment in a KBr sample cleaved from the same single crystal are plotted in Fig. 1(b). During loading similar jumps of the tip into the sample are detected at very similar loads as in the AFM experiment. A precise determination of the height of the jumps is difficult as they come close to the noise floor of the instrument. However, most of them seem smaller than 0.5 nm, and therefore, reflect single atomic-scale glide events similar to the AFM experiment. The nanoindentation results demonstrate that the total displacement of the tip into the sample is always less than 20 nm, i.e. clearly less than the estimated apex radius of the diamond tip. Consequently, the stress distribution in the crystal depends on the shape of only the tip apex which is comparable for the AFM tip and the indenter tip.

The comparability of the two experiments is impressively confirmed by the similarity of the load values at which atomic-scale glide nucleation is detected. Figure 1(b) also reveals a significant creep of about 5 nm while holding the indentation for 2 seconds at the maximum load of 10 μN. The unloading part of the curve shows a nonlinear character, and ends at a depth of 15 nm, indicating weak plastic recovery. The unloading part of the experiments will be discussed in more detail below.

Gaillard et al. have described the first pop-in event as the "brutal transition" between elastic and elasto-plastic behavior [14]. Comparable experiments by other groups also show dramatic pop-ins, a result of the creation of many dislocations in a very short time. Our experiments indicate that very sharp tips resolve the strain burst of the first pop-in into separated events of single dislocation activity. We have observed that the use of blunter tips in AFM-based indentation experiments results in pop-ins of many dislocations rather than the single-dislocation events described in this paper [7]. As a tentative conclusion we suggest that the observation of single dislocation nucleation in single crystal requires an indenting tip with an apex radius smaller than 30 nm.

Analysis of the surface topography after indentation

The AFM image in Fig. 2(a) shows the surface topography after an indentation into a KBr(100) surface up to a maximum load of 6 μN. The edge of the indent exhibits a nanometer-scale pile-up whose structure can not be resolved due to convolution with the tip apex. We believe that this pile-up in the immediate vicinity of the indent may be formed in part by displacement of molecules by the scratching tip [10] and in part by the processes of plasticity discussed below. Beyond the pile-up several terraces of monatomic height extend 100-200 nm from the indentation site. They have been formed in the course of the indentation. Their edges show preferential orientation along the [100] directions but also include irregularly curved parts. These terraces are formed by the glide and cross-slip of screw dislocations [11,8]. The monatomic steps indicate the paths taken by penetration points of screw dislocations with the surface. Two such penetration points are indicated by arrows. It should be noted that the localization and analysis of edge dislocation requires the detection of charges in Kelvin probe force microscopy and atomic-resolution imaging [12].

Geometry of dislocation activity

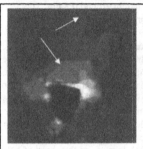

Figure 2: (a) AFM topography image of an indentation with a maximum load of 6 μN. Note the two screw dislocations indicated by arrows. (b) AFM image of the same area as in (a) showing a change of surface structure. The two screw dislocations moved to the right and downwards, respectively, and annihilated at the corner of an island indicated by the arrow. The process creates a new monatomic island (encircled). (frame size 220 nm)

In both AFM and nanoindentation experiments we detect sudden jumps of the indenting tip into the surface with a height corresponding to one atomic layer of the KBr(100) surface. In the AFM images recorded after indentation, we observe new monatomic steps in the vicinity of the indentation, mostly oriented along a [100] direction. The two observations can be explained by the activation of the primary glide system {110}<110> or the secondary glide system {100}<110>. Both leave a monatomic step along a [100] direction on the surface, in the case of the {110} glide only those planes which cut the surface under 45°. Note that in contrast to some other ionic crystals the secondary glide system is active in KBr at room temperature [13].

An attribution of the sudden jumps in the tip displacement to the breaking of a surface layer formed in atmosphere can be excluded for freshly cleaved KBr. The emergence of dislocations at the surface around the indent agrees with similar observations made on MgO and LiF samples [1,14].

The monatomic steps observed after indentation change direction, often in 90 angles. This has been observed first by Bassett using a decoration technique on bent KBr single crystals [15], later by other groups on NaCl also using decoration techniques [16], by AFM imaging [17], and finally on MgO by AFM [18]. The rectangular corners in surface steps have been attributed to cross-slip of screw dislocations into secondary glide systems. Similar to the detailed study of NaCl [16], the surface steps in our study mostly run along [100] directions, but have also components with a wide distribution of directions. We assume that the occurrence of cross-slips and the curvature of the path of screw dislocations reflect the strong stress gradients around the point-like indentation.

Reversal of plastic response for low maximum loads

The loading curve in Fig. 1(b) exhibits a strong curvature in the unloading curve for very low loads. This departure from the typical elastic curve indicates plastic recovery upon unloading. Note that unloading curves after indentation to loads higher than 100 μN (not shown here) result in an fully elastic behavior upon unloading, which allows for an analysis based on the methods introduced by Oliver and Pharr [9]. We conclude that the low number of dislocations introduced in these experiments at the onset of plasticity allow for a relaxation of the dislocation structure upon unloading. This idea is supported by an observations made during imaging of the dislocation structure by force microscopy. Processes of terrace formation through glide and cross-slip of dislocations are found to occur as late as 50 minutes after the original indentation. Such a sudden change of the surface structure is found in the transition from Fig. 2(a) to 2(b). A direct comparison of the new surface structure reveals the underlying process. Both screw dislocations have moved for several tens of nanometers, mostly along [100] directions. The path of the upper screw dislocation shows two instances of cross-slip. Finally, the two dislocations meet and annihilate at a pre-existing corner of a step edge, labeled by an arrow. It is interesting to note that both dislocations follow a curved path with a radius of about 10 nm just before they meet. The observed glide process illustrates how the movement of screw dislocations in a complete loop can create new monatomic terraces on the surface, here the encircled small island in the top part of Fig. 2(b). This is an interesting process of material transfer onto the surface based only on dislocation dynamics. Furthermore, it is of interest to note that the resulting new island is spatially disconnected from the indentation site. The annihilation of the two screw dislocations effectively lifts the surface in the vicinity of the indent, and therefore, would contribute to a nonlinear unloading curve like the one found in Fig. 1(b).

An interesting question is what activates the dislocation glide 20-50 minutes after the indentation, as observed in AFM. The strength of forces between tip and sample in the non-contact mode used for imaging is of the order of 1 nN, i.e. more than three orders of magnitude lower than the force used in the indentation. There is no indication that the beginning of the dislocation motion is correlated with the position of the scanning tip. We therefore conclude that the relaxation of stress through dislocation motion can be thermally activated as late as 20-50 minutes after indentation.

CONCLUSIONS

We have studied the onset of plasticity in KBr single crystals by means of nanoindentation with very sharp tips and the sharp tips used as probes in atomic force microscopy. Experiments employing both an instrumented nanoindenter and an atomic force microscope reveal atomic-scale yield events in the micro-Newton load regime. Each yield event is attributed to the glide of an active glide system by one lattice constant and to the nucleation of a single dislocation. Limiting the indentation to a maximum load of approximately 10 μN produces only a few dislocations, which have been observed to continue to glide and slip at times as late as 50 minutes after the indentation. Such dislocation mobility and the resulting partial reversal of plastic deformation are most likely related to the small number of dislocations created during the indentation process, which do not experience strong interactions with other dislocations.

REFERENCES

1. Y. Gaillard, C. Tromas, J. Woirgard: Quantitative analysis of dislocation pile-ups nucleated during nanoindentation in MgO, Acta Materialia 54 (2006) 1409–1417

2. W.W. Gerberich, W.M. Mook, M.J. Cordill, C.B. Carter, C.R. Perrey, J.V. Heberlein, S.L. Girshick: Reverse plasticity in single crystal silicon nanospheres, International Journal of Plasticity 21 (2005) 2391–2405

3. M.J. Cordill, M.D. Chambers, M.S. Lund, D.M. Hallman, C.R. Perrey, C.B. Carter, A. Bapat, U. Kortshagen, W.W. Gerberich: Plasticity responses in ultra-small confined cubes and films, Acta Materialia 54 (2006) 4515–4523

4. J.B. Pethica, R.S. Hutching, and W.C. Oliver: Phil. Mag. A 48 (1983) 593

5. J. Fraxedas, S. Garcia-Manyes, P. Gorostiza, and F. Sanz: Nanoindentation: Toward the sensing of atomic interactions, PNAS 99 (2002) 5228–5232

6. G.L.W. Cross, A. Schirmeisen, P. Grütter, and U.T. Dürig: Plasticity, healing and shakedown in sharp-asperity nanoindentation, Nature Material 5 (2006) 370

7. T. Filleter, S. Maier, R. Bennewitz: Atomic-scale yield and dislocation nucleation in KBr, Phys. Rev. B 73 (2006) 155433

8.. T. Filleter, R. Bennewitz: Nanometre-scale Plasticity of Cu(100), Nanotechnology 18 (2007) 044004

9. W.C. Oliver and G.M. Pharr: Measurement of hardness and elastic modulus by instrumented indentation, J. Mater. Res. 19, 3 (2004)

10. E. Gnecco, R. Bennewitz, and E. Meyer: Abrasive Wear on the Atomic Scale, Phys. Rev. Lett. 88 (2002) 215501

11. E. Carrasco, O. R. de la Fuente, M. Gonzalez, and J. Rojo: Dislocation cross slip and formation of terraces around nanoindentations in Au(001), Phys. Rev. B 68, 180102 (2003).

12. P. Egberts, T. Filleter, and R. Bennewitz: Kelvin Probe Force Microscopy of charged indentation-induced dislocation structures in KBr, Nanotechnology (June 2009)

13. W. Skrotzki, G. Frommeyer, P. Haasen: Plasticity of Polycrystalline Ionic Solids, phys. stat. sol. (a) 66, 219 (1981)

14. Y. Gaillard, C.Tromas and J.Woirgard: Pop-in phenomenon in MgO and LiF: observation of dislocation structures, Phil. Mag. Lett. 83, 553 (2003)

15. G.A. Bassett: The plasticity of alkali halide crystals, Acta Met. 7, 754 (1959)

16. F. Appel, U. Messerschmidt, V. Schmidt, O.V. Klyavin, A.V. Nikiforov: Slip Structures and Cross-slip of Screw Dislocations during the Deformation of NaCl Single Crystals at Low Temperatures, Materials Science and Engineering, 56 (1982) 211

17. W. F. Oele, J. W. J. Kerssemakers, and J. Th. M. De Hosson: In situ generation and atomic scale imaging of slip traces with atomic force microscopy, Rev. Sci. Instr. 68, 4492 (1997)

18. C. Tromas, J.C.Girard and J.Woirgard, Study by atomic force microscopy of elementary deformation mechanisms involved in low load indentations in MgO crystals: Phil. Mag. A 80, 2325 (2000)

Mater. Res. Soc. Symp. Proc. Vol. 1185 © 2009 Materials Research Society   1185-II07-10

Stress-strain behavior of individual electrospun polymer fibers using combination AFM and SEM

Fei Hang[1], Dun Lu[1], Shuang Wu Li[1] and Asa H. Barber[1]
[1]Centre for Materials Research & School of Engineering and Materials Science, Queen Mary University of London, Mile End Road, London E1 4NS, UK

ABSTRACT

Tensile deformation of individual electrospun polyvinyl alcohol (PVA) nanofibres was performed using a novel combination atomic force microscope (AFM)- scanning electron microscope (SEM) technique. The AFM was used to provide manipulation and mechanical testing of individual PVA nanofibers while the SEM was used to observe the deformation process. Resultant stress-strain curves show how the elastic modulus shows comparable, or even slightly increased, values to isotropic films. In addition, the electrospun fibers were tested to failure to measure their tensile strength.

INTRODUCTION

Polymers are typically processed into fibers especially when improvements in mechanical properties are required. Manufacturing methods such as melt spinning [1] and gel spinning [2] induce significant structural anisotropy with alignment of the polymer chains along the principle fiber axis. In some cases, the mechanical properties are considerable such as an elastic modulus of the order of many tens of GPa. Electrospinning is a historical process which has been recently revived for the production of polymer fibers [3] with diameters extending to the nanometer range [4]. The electrospinning process is potentially interesting as fibers are drawn from a polymer solution through an electric field to form solid fibers. The action of the electric field and the drawing process may induce structural anisotropy, thus improving the mechanical properties of the polymer.

Mechanical testing of electrospun polymer fibers can be achieved through conventional techniques or the application of novel techniques to probe individual fiber mechanical performance. Conventional larger scale testing methods are expected to be inferior to mechanical testing of individual fibers due to the agglomeration of fibers which need to be tested together, resulting in external loading causing potential inter-fiber sliding as well as mechanical performance being averaged over the total number of fibers within the test. Tensile testing of individual electrospun polyethylene oxide fibers has been achieved using optical microscopy to observe the mechanical testing of a fiber of relatively large diameter [5]. One end of the fiber ends was attached to a micromanipulator whereas the other free end was glued to a small beam. Translation of the micromanipulator caused deformation of the fiber, with the beam bending used to quantify the force applied to the fiber. The technique was somewhat limited by manipulation of the fiber being observed by an optical method, indicating that electrospun fibers with diameters of a few hundred nanometers of less could not be resolved optically. Improvements in mechanical testing were achieved using a microelectromechanical device to apply tensile forces to individual electrospun fibers with diameters ranging from 300-600nm [6]. Resultant force-displacement curves showed a highly elastic response for electrospun polyacrylonitrile fibers but challenges in both the mechanical testing of fibers of smaller diameters and high force resolution still need to be overcome.

This paper uses a novel method of scanning electron microscopy (SEM) to allow manipulation of relatively small diameter electrospun nanofibres and high force resolution mechanical testing using atomic force microscopy (AFM). The AFM is present within the SEM chamber to allow full in situ visualization during the AFM mechanical testing.

EXPERIMENT

SAMPLE PREPARATION

Electrospinning fibers was achieved by first dissolving PVA powders ($M_w =$ 85,000-146,000 g.mol^{-1}, DH=99%; Sigma-Aldrich, UK) in distilled water at 80 °C and gently stirred for 3 hours to give a polymer solution. The polymer concentrations varied from 7-9.5 wt/v% in the solution. Electrospinning experiments were carried out according to our previous works [7]. Generally, this involved introducing the polymer solution into a metal tipped syringe and applying a voltage between the end of the tip and a silicon grounded substrate until ejection of fibers from the end of the needle was achieved at approximately 12-15kV. The mechanism of this electrospinning process has been described elsewhere [3]. A mat of electrospun fibers on the substrate was then transferred into the SEM chamber.

MECHANICAL TESTING

A custom built AFM using commercially available components (Attocube GmbH, Ger,) was integrated into the chamber of a high resolution SEM (Quanta 3D, FEI Company, EU). The SEM provided visualization of the manipulation process but care was taken to minimize the effect of the electron beam on the sample by using a long working distance (15mm), low accelerating voltage (5kV) and low electron beam current (40pA). The custom built AFM within the SEM chamber was then used to manipulate and attach the end of an electrospun PVA fiber to the apex of the AFM probe. This was achieved by translating the end of the AFM probe into a droplet of glue (Poxipol, Arg.) present within the chamber to allow transfer of some of the glue to the end of the probe. The AFM probe was then moved towards a contact with an individual electrospun fiber protruding from the spun fiber mat. The glue was then allowed to fully solidify, generally within a time frame of 10 minutes, to ensure that the PVA fiber was securely attached to the end of the AFM probe. The electron beam of the SEM was then focussed onto a point on the fiber approximately 10-20μm away from the fiber-glue contact point in order to melt through the end of the PVA fiber. This focus was typically at magnifications of around x1,000,000 with an increased accelerating voltage to assist in the melting process. The result of the manipulation was to give a separated individual electrospun PVA fiber attached to the end of the AFM probe. Fresh glue was then introduced into the chamber

and the free end of the PVA fiber pushed into the semi-liquid glue, with a resultant
typical setup shown in Figure 1.

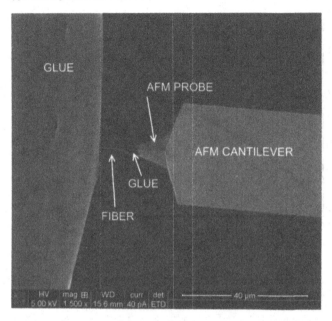

Figure 1. SEM secondary electron image of an individual PVA fiber attached between a
glue droplet (left) and a second smaller glue droplet (right) present on the end of an AFM
probe. Translation of the AFM probe to the right causes tensile deformation of the
electrospun fiber, with the force recorded during this deformation from bending of the
AFM cantilever.

After solidification of the glue, the AFM probe was translated such that a tensile
strain was applied to the PVA fiber until failure of the fiber occurred. The rate of
displacement was approximately $2\mu m.s^{-1}$. The SEM facilitated direct observation of the
failure process during the deformation of the fiber. Qualitative observations indicate that
the fiber failed in an almost ductile manner. The force applied to the PVA fiber was
accurately determined from an interferometer setup monitoring the AFM cantilever
deflection. Strain was simply calculated from the piezo-positioner movement during the
test minus the cantilever bending. Resultant stress-strain curves for 3 PVA fibers are
shown in Figure 2.

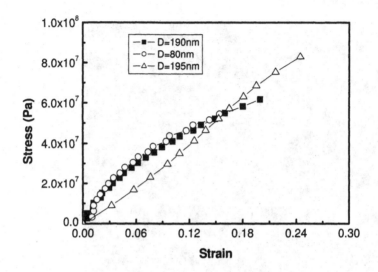

Figure 2. Tensile stress-strain curves for 3 different electrospun PVA fibers with different diameters (D).

DISCUSSION

The results in Figure 2 indicate that the deformation of the electrospun PVA is initially linear but with increasing non-linearity indicating viscoelastic behavior for at least two sets of data (the squares and circles). All of the stress-strain curves indicate considerable linear behavior compared with typical PVA materials which exhibit considerable plastic deformation. The origin of this stress-strain behavior in electrospun PVA fibers is unknown but may be due to the difficulty in plastic deformation at reduced length scales examined in this paper. For the tested fiber with a diameter of 195nm, the set of data appears to be fairly linear which suggests more localized failure. The elastic modulus for the linear data (triangles) is 385MPa which suggests that the mechanical properties of the electrospun PVA is only marginally higher than bulk isotropic PVA film, measured using conventional Instron tensile testing and found to be 211±105MPa. However, the initial elastic modulus of the other two data sets varies from 605-740MPa in the initial regime of up to 3% strain indicating that the fibers have potential structural anisotropy along the tested fiber direction. The results present here would therefore indicate that the electrospun fibers have an elastic modulus which is at least comparable to bulk PVA values. It is worth noting that the PVA fibers can swell under the influence of water, with dry fibers giving elastic modulus values of around 3GPa [8]. The fibers and films in this paper are not dried which suggests that water is still bound within the PVA fibers despite conducting the tests in the vacuum chamber of an SEM.

CONCLUSIONS

A combined AFM-SEM was used to mechanically testing individual electrospun fibers of PVA. This technique is particularly effective at measuring the mechanical properties of individual nanomaterials as demonstrated using the electrospun fibers in this work. The mechanical properties of individual electrospun PVA fibers, as with macroscopic samples, were characterized from stress-strain curves obtained from using combination AFM-SEM. In addition, the strength and elastic modulus of the PVA fibers are comparable to bulk PVA behavior although there is some indication of some improvement in the PVA from the electrospinning process, perhaps due to an increase in molecular alignment.

ACKNOWLEDGMENTS

The authors would like to thank the Engineering and Physical Science Research Council UK (grant EP/E039928/1) for financial assistance, NanoForce Technology for access to electrospinning facilities and the NanoVision Centre at QMUL for microscopy facilities.

REFERENCES

1. J. R. Dees, J. E. Spruiell, J. Appl. Polym. Sci. 18 (1974), 1053
2. P. Smith, P. J. Lemstra, 15 (1980), 505
3. J. Doshi, D. H. Reneker, 35 (1995), 151
4. D. H. Reneker, I. Chun, Nanotechnology 7 (1996), 216
5. E. P. S. Tan; C. N. Goh, C. H. Cow, C. T. Lim, Appl. Phys. Lett. 86 (2005), 073115
6. M. Naraghi, I. Chasiotis, H. Kahn, Y. Wen, Y. Dzenis, Rev. Sci. Instr. 78 (2007), 085108
7. W. Wang, P. Ciselli, E. Kuznetsov, T. Peijs, A. H. Barber, Phil. Trans. R. Soc. A 366 (2008), 1613
8. M. K. Shin, S. I. Kim, S. J. Kim, S.-K. Kim, H. Lee, Appl. Phys. Lett. 88 (2006), 193901

Mater. Res. Soc. Symp. Proc. Vol. 1185 © 2009 Materials Research Society   1185-II10-01

Measuring the Glass-Transition Temperature in Electrospun Polyvinyl Alcohol Fibers Using AFM Nanomechanical Bending Tests

Wei Wang[1] and Asa H Barber[1]
[1]Centre for Materials Research & School of Engineering and Materials Science, Queen Mary University of London, Mile End Road, London E1 4NS, UK

ABSTRACT

The glass transition of individual electrospun PVA fibers was found using an AFM fiber bending technique within a heated chamber. A considerably loss in the measured elastic modulus was observed with temperature when the glass transition temperature was reached. The glass transition temperature was observed to decrease as the electrospun PVA fiber diameter decreased, indicating diameter dependent enhanced polymer chain mobility.

INTRODUCTION

Electrospinning is becoming a common method for the production of continuous polymer fibers [1]. The geometry of the resultant electrospun polymer fibers is critically linked to the electrospinning parameters [2]. In particular, the diameter of the spun fibers can be reduced to lengths approaching tens of nanometers but the effects of variation in size on physical properties has yet to be accurately determined. Understanding the variation of physical properties with fiber diameter needs to be accurately determined at a specific diameter but electrospinning does not produce fibers with mono-disperse diameters. Therefore, conventional techniques which require large amounts of electrospun fibers are insufficient as the corresponding properties over the range of fiber diameters will be examined.

Some literature from our group [3] has recently examined the thermo-mechanical properties of electrospun polyethylene oxide fibers using atomic force microscopy (AFM). The advantage of this technique was to evaluate mechanical properties of individual electrospun fibers of a specific diameter and observe how the mechanical properties vary up to the melting temperature. In this work we extend the approach of nanomechanical testing to evaluate the glass transition temperature of electrospun polyvinyl alcohol (PV) fibers over a range of fiber diameters.

EXPERIMENT

Electrospinning was performed according to previous methods. Briefly, electrospinning introduces a polymer solution to a syringe with a metal needle tip. A voltage is applied between the syringe and a grounded collector substrate. The field strength at the end of the needle tip is considerably higher than in other regions within the syringe and, at a critical voltage, the build up of change within the solution causes repulsion within the solution and an expulsion of the solution from the syringe towards the collector. In this work, PVA (Sigma, UK, M_w= 85,000-146,000) was dissolved in water at 80°C such that polymer solutions with a PVA wt/vol.% ranging from 6-13% could be prepared in order to spin fibers of different diameters. A polymer film with thickness of many hundreds of microns was also cast from the same solution and is designated as a bulk sample. Fibers of PVA were electrospun at applied voltages of around 12kV and collected onto a patterned silicon substrate with a number of trenches within the substrate. This collector grid was transferred to an AFM with combination heating stage (NTegra, NT-MDT, Rus.) for thermo-mechanical testing. The AFM probe and sample are contained within the same heating chamber to ensure that the temperature of both the sample and AFM probe are the same. The grid was first imaged in a semi-contact mode in order to locate individual electrospun fibers bridging the trenches in the grid. The mid-point of a bridging electrospun fiber was selected and the AFM probe pushed into the mid-point using the z-piezo positioner of the AFM setup in order to cause bending of the bridging fiber. The force required to bend the electrospun fiber was found by recording the AFM cantilever bending and converting to force using the Sader method [4]. Deflection of the bending electrospun fiber was simply found by taking the z-piezo positioner movement minus the AFM cantilever bending. A typical force verses z-piezo displacement curve is shown in Figure 1 below. Note that the curve is linear in Figure 1, indicating that elastic deformation of the fiber was achieved. Possible slippage of yielding of the fiber did not occur due to this linear relationship.

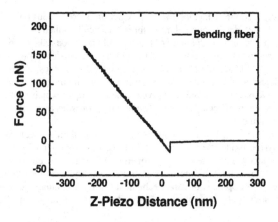

Figure 1. Plot of the force applied to the midpoint of a bridging individual electrospun PVA fiber with z-piezo displacement distance.

The bending of individual electrospun PVA fibers was monitored as a function of temperature by heating the samples from room temperature to 100°C using the heating chamber of the AFM setup.

RESULTS AND DISCUSSION

The bending elastic modulus (E_b) of individual electrospun PVA fibers can be evaluated by considering the fiber as a clamped elastic beam using [5]:

$$E_b = \frac{F}{\delta}\frac{L^3}{192I}$$ Equation 1

Where F is the applied normal force, δ is the deflection at the mid-point of the freely suspended fiber length L, I is the second moment of inertia defined for a cylindrical beam with diameter D_f as $I = \pi D_f^4/64$. The L and D_f dimensions were taken from AFM images and δ was obtained from the z-piezo displacement (such as shown in Figure 1) minus the AFM canilever bending. The bending modulus is equal to the elastic modulus of the PVA fiber (E_f) when the length-to-diameter ratio is greater than 16, as achieved in this work, otherwise significant shear forces can develop during the bending process, and for small deflections (less than the fiber radius) which are used in this paper. A typical plot of the calculated elastic modulus with temperature is shown in Figure 2.

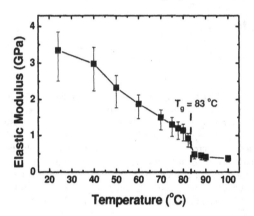

Figure 2. Plot of the elastic modulus of an individual PVA fiber, measured using AFM bending, against temperature with a highlighted glass transition temperature of 83°C.

The elastic modulus is seen to fall with temperature with a notably rapid drop in elastic modulus to around 500MPa. This temperature corresponds approximately to the expected glass transition temperature (T_g) for PVA. T_g values for 3 different PVA fiber diameters and a bulk PVA film, measured using conventional differential scanning calorimetry, are shown in Figure 3 below.

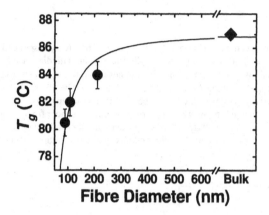

Figure 3. Plot of glass transition temperature with fiber diameter for electrospun PVA fibers (circles) and a bulk PVA film (diamond).

The measured T_g values for the electrospun PVA fibers are comparable to that of the bulk PVA film. This confirms that the AFM bending technique is sufficient to resolve glass transitions for individual polymer fibers. The glass transition is seen to decrease as the fiber diameter decreases in Figure 3. Previous works have investigated the T_g in polymer films and also observed a similar effect of a T_g suppression as the film thickness decreases [6]. Increased polymer chain mobility at a polymer surface has been proposed as a mechanism to support these observations [7]. Generally, T_g is expected to decrease as the polymer chain mobility increases. The observed T_g depression with reduced diameter would therefore indicate that the relatively large fiber surface relatively to fiber volume gives a enhanced polymer mobility at the surface and, thus, lower T_g values when compared to the bulk PVA film.

CONCLUSIONS

The glass transition temperature of electrospun PVA fibers were found by bending individual fibers using AFM. A sudden drop in the elastic modulus of the PVA fiber within a temperature range of 80-85°C was shown to correspond to the glass transition of the PVA, highlighting the applicability of AFM mechanical measurements to determine thermal transitions at the nanoscale. A depression in the T_g with decreasing fiber diameter was observed, suggesting an increased polymer chain mobility at the surface of electrospun PVA fibers.

ACKNOWLEDGMENTS

This work was partially supported by the University of London Central Research Fund. Authors acknowledge NanoForce Ltd. and the NanoVision Centre at QMUL for using electrospinning and AFM facilities. Thanks are also given to Dr. M. Tian at NT-MDT, Europe BV for discussions of AFM measurements.

REFERENCES

1. D. Li, Y. Xia, Adv. Mater. 16 (2004), 1151.
2. S. V. Fridrikh, J. H. Yu, M. P. Brenner; G. C. Rutledge, Phys. Rev. Lett. 90 (2003), 144502.
3. W. Wang, A. J. Bushby, A. H. Barber, Appl. Phys. Lett. 93 (2008), 201907.
4. J. E. Sader, J. W. M. Chon; P. Mulvaney, Rev. Sci. Instr. 70 (1999), 3967.
5. J. M. Gere; S. P. Timoshenko, "Mechanics of Materials." 4th ed. 1997, Boston, MA: PWS Pub Co.
6. J. L. Keddie, R. A. L. Jones, R. A. Cory, Europhys. Lett. 27 (1994), 59.
7. J. A. Forrest, J. Mattsson, Phys. Rev. E 61 (2000), R53

Mater. Res. Soc. Symp. Proc. Vol. 1185 © 2009 Materials Research Society   1185-II11-03

In-Situ TEM Investigation of Deformation Behavior of Metallic Glass Pillars

C.Q. Chen, Y.T. Pei, J.Th.M. De Hosson*

Department of Applied Physics, Materials Innovation Institute M2i, University of Groningen, Nijenborgh 4, 9747 AG Groningen, The Netherlands

ABSTRACT

We show results of in situ TEM (transmission electron microscope) quantitative investigations on the compression behaviors of amorphous micropillars fabricated by focused ion beam from $Cu_{47}Ti_{33}Zr_{11}Ni_6Sn_2Si_1$ metallic glass (MG) ribbon. Pillars with well defined gauge sections and tip diameter ranging from 100 nm to 640 nm are studied. Quantitative compression tests were performed by a recently developed Picoindenter TEM holder, with the evolution of individual shear bands monitored in real time in TEM. It is found that the deformation of the MG pillars at the present size domain is still dominated by discrete shear banding as demonstrated by intermittent events in the load-displacement curves. However, the frequency, amplitude and distribution of these shear banding events are clearly size-dependent at submicrometer scale, leading to an apparently transition in deformation mode from highly localized inhomogeneous deformation to less localized and more distributed deformation with decreasing pillars diameter. Deformation of a 105 nm diameter pillar having rounded tips is characterized with fully homogeneous bulge at the initial stage of deformation, indicating prompting effect of multi-axial stress state on transition to fully homogeneous deformation.

INTRODUCTION

Despite the high yield strength, monolithic metallic glass suffers from highly localized shear deformation at ambient temperature and therefore exhibit very limited ductility [1-3], with the process of initiation and evolution of shear bands still puzzling [3]. Study of the deformation behavior of small volume metallic glass is an interesting route for the exploration of the initiation and evolution of individual shear bands and it has attracted a rapid increasing interest over recent years [4,5,6,7,8,9,10,11]. It also has practical significance on the design of recently booming metallic glass based composites [12] or multilayers [13], which is effective in improving the ductility but needs understanding in shear localization that is constrained by size. Various hints in deformation mechanism of small volume metallic glasses appeared suggesting either improved ductility [4,5,11] or increased yield strength at small scales [6,7,8,9,10]. However, due to technical difficulties, the information obtained is limited by a lack of either quantitative stress-

* Corresponding author. Tel.: +31-50-363 4898; Fax: +31-50-363 4881
E-mail address: j.t.m.de.hosson@rug.nl

strain information [4] or a capability of monitoring the evolving deformation structure [5,6,7,8,9,10], as well as by the limited range of examined specimen sizes. We report here quantitative in-situ TEM microcompression of metallic glass pillars with diameters ranging from 640 nm down to 105 nm, with new insights into size-effects on deformation of metallic glasses obtained.

EXPERIMENTAL DETAILS

Micropillars were cut by focused ion beam (FIB) from two metallic glass ribbons of Cu47Ti33Zr11Ni6Sn2Si1 prepared by melt spinning. By careful design of the annular FIB milling procedure while reducing the ion beam current from 100 pA for rough cutting to exceptionally low current of 20 pA for final shaping, pillars with well defined gauge section and tip diameter ranging from 93 nm to 645 nm were successfully fabricated. The pillars were shaped to have a rather small taper angle between 2.0-3.5°, and importantly well-defined gauge lengths ranging from 1.2-2.0 µm, and aspect ratio from 3 to 8. In situ TEM compression experiments were performed using a recently developed Hysitron picoindenter TEM holder (Hysitron Inc., Minneapolis, MN) equipped on JEOL 2010F TEM, with a diamond flat punch of 2 µm diameter. The picoindenter is integrated with a miniature capacitive load–displacement transducer permitting high resolution load and displacement measurements (resolution of ~0.3µN in load, ~1 nm in displacement). Rapid instrument response and data acquisition rates (up to the order of 10^4/s, the controller operating in a continuous loop and sampling data at 80 kHz) allows discrete flow events well resolved.

The compression experiments were performed in two different control modes, i.e. displacement rate control that exhibits a greater sensitivity to transient load drops, and load rate control that has advantage in evaluating sudden displacement jumps (shear band offset), with corresponding displacement or load rate programmed such that a nominal strain rate of ~10^{-2}/s is used. The engineering stress in the curves shown in Fig. 2 is approximated by: $4P/[\pi(D+2u\tan\beta)^2]$ where P is the load and D tip diameter, β taper angel, u measured tip displacement.

DISCUSSION

In-situ TEM Microcompression of 640 nm tip-diameter pillar

Fig. 1 shows the compression of a Cu-based pillar with 640 nm tip diameter. The pillar shows elastic followed by jerky type deformation behavior with transient shear banding events registered with displacement bursts in the curves. Shear bands initiated sequentially from the top of the pillar due to combined effect of the slight tapering and the initial imperfect pillar-punch contact. The successive initiation of shear bands cause respectively a number of burst events in the load-displacement curve. The local plastic deformation makes effective diameter of the top part of the pillar increase, resulting in a load increase upon further loading. Subsequently, the pillar deforms with a fast major shear process carrying a large shear displacement. Markedly, after the major shear process the pillar is not fractured, upon further loading the load increased very fast and almost reached the previous value. Fracture was observed only when the tip was manually retracted after unloading since the tip adhered strongly to the punch, and the top part of the pillar is sheared off from the lower part along a preexisting shear band produced during the compression. The fracture tolerance and large engineering plastic strain obtained can be reasonably explained by an "extrinsic" size effect [7], since the critical shear offset at which failure happens (typically taken as 10-20 micrometers) [14,15] can not be reached in such small pillars.

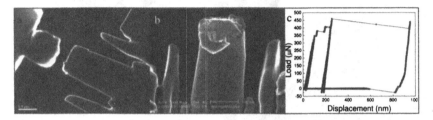

Figure 1 After-deformation TEM image (a) and SEM image (b) showing the deformed structure of a ⌀640 nm Cu-based metallic glass pillar compressed under load-control. The load-displacement curve is shown in (c), which shows occurrence of a major shear band carrying majority of deformation registered with a large displacement burst in the curve.

In-situ TEM Microcompression of 400 nm tip-diameter pillar.

Figure 2 shows compression of a pillar with 400 nm tip diameter. Compared to the appearance of the major shear band in the 640 nm pillar, the deformation of this thinner pillar is accommodated by a much larger number of shear banding events. With the plastic deformation proceeds, shear bands were triggered successively causing a large number of intermittent burst events in the load-displacement curve. Remarkably, none of these shear bands run over a long distance upon initiation, instead they initiate with a small shear offset (~20 nm). It seems that the shear bands propagation in the 400 nm diameter pillar is more remarkably suppressed by size effects. With a propagating shear band arrested, a new shear process will be triggered upon further increased load and elastically accumulated energy. This mechanism results in a frequent shear banding suppression and reinitiation associated with the intermittent events in the curve. The reinitiation can occur either on the preexisting one or in a neighboring virgin region. The post-deformation SEM image shows multiple shear bands distributed rather uniformly through out the gauge section (Figure 2c). Compared to the 640 nm pillar, it seems there is an apparent change of deformation mode, i.e., from highly localized and strongly inhomogenous deformation at 640 nm diameter to less localized and more homogeneous deformation at 400 nm diameter.

101

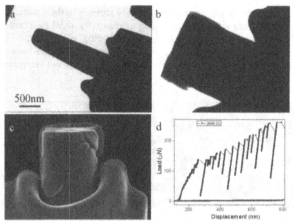

Figure 2. TEM images showing the structure of a ⌀400 nm pillar before (a) and after (b) deformation. (c) SEM image showing the post-deformation surface morphology with multiple shear bands clearly observable. The load-displacement curve is shown in (d), with the measured yield stress also indicated.

In-situ TEM Microcompression of 105 nm tip-diameter pillar.

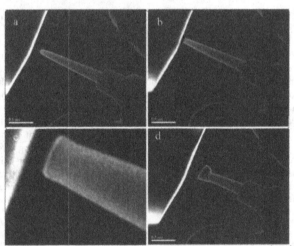

Figure 3 Micrompression of a 105 nm tip diameter pillar. TEM images showing the structure of a ⌀440 nm pillar before (a) and after (b) initial deformation; (c) close up of the deformed tip region showing a rather homogeneous bulge; (d) with deformation continues, shear banding takes over and becomes the major deformation mode.

Microcompression of a much thinner pillar with tip diameter of 105 nm shows that the deformation is still dominated by intermittent flow with abundant jerky type events happening. However, interestingly, it is observed that for this extremely thin pillar having a rounded tip (inevitable due to FIB milling procedure for such a small diameter), the initial stage of deformation is characterized by a fully homogeneous bulge (mushrooming) morphology (Figure 3c), indicative a kind of deformation mode transition from inhomogeneous to fully homogeneous. However, while deformation proceeds subsequently shear bands occur and become the major deformation mode. Presumably, the switch of deformation mode to fully homogeneous at the tip is due to a combined effect of the extremely small size and stress triaxiality induced by the rounded geometry, suggesting a prompting effect of stress-triaxiality on appearance of fully homogeneous deformation.

CONCLUSIONS

We show several striking new results of in situ TEM quantitative investigations on the compression behaviors of amorphous micropillars fabricated by focused ion beam from $Cu_{47}Ti_{33}Zr_{11}Ni_6Sn_2Si_1$ metallic glass ribbon. Deformation of pillars with three typical diameters of 640, 400 and 105 nm is highlighted in this report. It is found that the deformation of the MG pillars at the present size domain is still dominated by intermittent plastic flow, which is accommodated by discrete shear banding events. However, the frequency, amplitude and distribution of these shear banding events are strongly size-dependent at submicromenter scale, leading to an apparently transition in deformation mode from highly inhomogeneous to rather homogeneous with decreasing pillars diameter. Deformation of the 105 nm diameter pillar having rounded tips is characterized with fully homogeneous bulge at the early stages of deformation, indicating prompting effect of multi-axial stress state on deformation mode transition.

ACKNOWLEDGMENTS

The authors acknowledge financial support from the Netherlands Materials Innovation Institute (M2i) and the Foundation for Fundamental Research on Matter (FOM-Utrecht), the Netherlands. We thank Dr. Frans D. Tichelaar, Dr. Vasyl. L. Svechniko, Dr. Paul Alkemade in the Technology University of Delft for help in FIB fabrication of the specimens, and thank Dr. Paul M. Bronsveld and Dr. Vasek Ocelik in University of Groningen for providing the metallic glass ribbons.

REFERENCES

1. Greer AL. Science 1995;267:1947.
2. Inoue A. Acta Mater 2000:48:279.
3. Schuh CA, Hufnagel TC, Ramamurty U. Acta Mater 2007;55:4067.
4. H. Guo, P. F. Yan, Y. B. Wang, J. Tan, Z. F. Zhang, M. L. Sui, E. MA. Nat. Mater. 6, 735 (2007).
5. Q. Zheng, S. Cheng, J. H. Strader, E. Ma and J. Xu, Scr. Mater. 56, 161 (2007).
6. C. J. Lee, J. C. Huang, T. G. Nieh. Appl. Phys. Lett. 91, 161913 (2007)
7 S. Cheng, X L Wang, P. K. Liaw. Appl. phys. Lett. 91, 201917 (2007)

8 B. E . Schuster , Q. Wei, M.H. Ervin, S. Q. Hruszkewycz , M. K. Miller, T.C. Hufnagel, K. T. Ramesh. Scr. Mater. 57, 517 (2007)

9 Z. W. Shan, J. Li, Y. Q. Cheng, A. M. Minor, S. A. Syed Asif, O. L. Warren, and E. Ma. Phys. Rev. B 77, 155419 (2008)

10 Y. H. Lai, C. J. Lee, Y. T. Cheng, H. S. Chou, H. M. Chen, X. H. Du, C. I. Chang, J. C. Huang, S. R. Jain, J. S. C. Jang, T. G. Nieh. Scr. Mater, 58, 890, (2008).

11 C. A. Volkert, A. Donohue, and F. Spaepen. J Appl Phys. 103, 083539 (2008).

12. C. C. Hays, C. P. Kim, W. L. Johnson. Phys. Rev. Lett. 84,2901 (2000).

13 Y. M. Wang,J. Li, A. V. Hamza, and T. W. Barbee, Jr. Proc. Natl. Acad. Sci. 104, 11155 (2007).

14. Conner RD, Johnson WL, Paton NE, Nix WD. J Appl Phys 2003;94:904.

15. Ravichandran G, Molinari A. Acta Mater 2005;53:4087.

Mater. Res. Soc. Symp. Proc. Vol. 1185 © 2009 Materials Research Society   1185-II03-15

Could Life Originate Between Mica Sheets? Mechanochemical Biomolecular Synthesis and the Origins of Life

Helen Greenwood Hansma
Department of Physics, University of California
Santa Barbara, CA 93106

ABSTRACT

The materials properties of mica have surprising similarities to those of living systems. The mica hypothesis is that life could have originated between mica sheets, which provide stable compartments, mechanical energy for bond formation, and the isolation needed for Darwinian evolution. Mechanical energy is produced by the movement of mica sheets, in response to forces such as ocean currents or temperature changes. The energy of a carbon-carbon bond at room temperature is comparable to a mechanical force of 6 nanoNewtons (nN) moving a distance of 100 picometers. Mica's movements may have facilitated mechanochemistry, resulting in the synthesis of prebiotic organic molecules. Furthermore, mica's movements may have facilitated the earliest cell divisions, at a later stage of life's origins. Mica's movements, pressing on lipid vesicles containing proto-cellular macromolecules, might have facilitated the blebbing off of 'daughter' protocells. This blebbing-off process has been observed recently in wall-less L-form bacteria and is proposed to be a remnant of the earliest cell divisions (Leaver, et al. *Nature* **457**, 849 (2009).

INTRODUCTION

The hypothesis of this paper is that mica is in many ways an ideal environment for the origins of life. One potential advantage of mica is that it could provide an endless source of mechanical energy for synthesizing the many covalent bonds needed for even the simplest life.

The scenario might be something like this: Simple molecules bind to the edges and surfaces of the sheets in a mica book. The sheets move up and down in response to heating and cooling and water flow. These movements squeeze and stretch the molecules with enough force to make and break covalent bonds between them. Larger molecules are formed, including amino acids and other common monomers in the bio-polymers of life. Impurities in the mica lattice, such as iron, serve as reaction centers that bias the mechano-chemical reactions in favor of some molecular products over others. As molecular products accumulate, they diffuse between mica sheets and orient on the mica lattice. Their orientation is determined by the 0.5-nanometer periodicity of the mica lattice and its electronegativity. These constraints of the mica lattice bias the molecular binding, favoring isomers of the same handedness: either L- or D-isomers bind nicely next to each other on the mica lattice, while mixtures of L- and D-isomers do not bind as close to each other. Closely spaced molecules join, by losing a water molecule, and polymers form. These polymers are especially stable in dry spaces between mica sheets, where mass action favors polymerization over hydrolysis.

As time passes, polymers of different types accumulate between mica sheets. Different polymers form and accumulate between different mica sheets, due to stochastic processes. In some regions, RNA worlds may develop. In other regions, peptides may predominate, or lipids. Figure 1 is a sketch of the hypothesized system at this polymer-rich stage in prebiotic evolution.

Over time, complexity will increase. Self-replicating ribozymes from nearby regions will join to form larger ribozymes capable of storing and transmitting more genetic information. Ribopeptides may form from the interactions of RNA and peptides. Lipid bilayers self-assemble on and between mica sheets. These lipid bilayers periodically encapsulate collections of molecules on the mica sheets or their edges. Occasionally the lipid bilayers encapsulate a viable collection of molecules, capable of self-replication, metabolism, and some form of cell division. At least one of these collections of molecules survived and evolved into life as we know it today.

For delightful books on the origins of life, see Hazen,[1] for the research and the people who did it, and Dyson,[2] for a toy model of how prebiotic molecules might have evolved.

Figure 1. The mica hypothesis for the origins of life. A sketch of mica sheets under water, with 'molecules' of various sizes and conformations between the sheets. Mica sheets are 1-nm thick. The sketch shows the mica separated into layers as thin as 3-4 sheets (3-4 nm), but it is more realistic to propose that the thinnest layers are hundreds of sheets thick, to have larger spring constants and more robust compartments.

Mica

Mica is a layered mineral with 1-nm-thick sheets bridged by potassium ions (K^+). These sheets are composed mostly of silicon (Si), oxygen (O) and aluminum (Al). Each 1-mn-thick sheet has 3 layers. The top and bottom layers of the sheets are a hexagonal net of mostly Si and O. The middle layer, sandwiched between the sheets' surfaces, is an octagonal layer of mostly Al and O. This Al-O layer has hydroxyl groups that are recessed slightly below the Si-O surface. These recessed hydroxyl groups are hexagonally spaced and 0.5 nm apart. Half of the recessed hydroxyls are ionized, giving a surface charge of 2 negative charges per square nanometer, in the absence of counterions. In unsplit mica, potassium ions bridge the recessed hydroxyls in adjacent sheets and hold the sheets together.[3]

Mica has an affinity for biomacromolecules of many types. Mica's affinity for DNA correlates with the ionic radius of the inorganic cation used to bind the DNA to the mica.[4]

Mica has a clay-like layered chemical structure on the scale of its crystal unit cell. Unlike clays, which have micron- or sub-micron-sized sheets, mica's crystalline sheets extend for millimeters, centimeters, and more.

HYPOTHESIS

Mica has the possibility of transducing solar or other heat energy and kinetic energy from water movements into mechanical energy for stretching and compressing molecules between mica sheets. A beautiful advantage of mechanical energy for life's origins is that it is endlessly available between mica sheets, due to water movements at mica's edges and temperature changes that may create heat pumps in mica's bubble defects.

Orientation and compression to inter-atomic distances should be able to form covalent bonds of prebiotic monomers and polymers. Polymerization of monomers is hypothesized to occur when monomers orient in adjacent sites on the mica lattice. The 0.5-nm periodicity of the anionic mica lattice is the same as the spacing between nucleotides in single-stranded nucleic acids and between amino acids in extended β-sheet structures. Inorganic cations are proposed to bridge the monomers and polymers to the mica.

These flows, expansions, and contractions of fluids exert mechanical energy in a cyclic fashion on whatever is present between the mica sheets. Forces and pressures would be expected to vary over many orders of magnitude, depending on the thickness and area of mica sheets, the temperature changes, the sizes of air bubbles, and other factors.

As shown in Figure 2, two types of mechanical energy are envisioned for mica. In the first type, fluid flows in and out of the spaces between mica sheets, with the ebb and flow of water currents from prebiotic oceans or lakes. In the second type, fluid and air bubbles between mica sheets expand and contract during the earth's daily cycles of heating and cooling. This second type is a hot-air engine or heat pump.

Mechanical Energy from Mica?

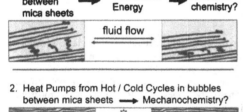

Figure 2. Sketches of two possible sources of mechanical energy from mica, for synthesis of prebiotic molecules. Top panel shows fluid flows at the edges of mica sheets that would stretch and compress molecules. Capillary forces would probably be the most likely forces between mica sheets with the smallest separations. Bottom panel shows how heat pumps or hot air engines might operate in the bubble defects in mica.

Heat pumps have been proposed as a possible energy source for the origins of life[5] and are hypothesized to form in mica defects. Even the highest grade of mica has visible defects that appear to be flattened air bubbles between mica sheets. These bubble defects appear as roughly circular areas, typically several millimeters in diameter. These bubbles between mica sheets may expand and contract by a few picometers or more in response to solar heating and night-time cooling. Thus mica may function as a heat pump. This source of mechanical energy from mica is more problematic than the first source in Figure 2, arising from fluid flows between mica sheets. Heat pumps raise questions such as how small precursor molecules got between the mica sheets. The first panel in Figure 2 shows a mechanism that seems highly probable.

In the earliest stages of prebiotic molecular synthesis, the up-and-down motions of mica sheets might bring small molecules close enough together to enter the attractive regime of the energy-vs.-distance profile (Figure 3). At the earliest stages, monomeric molecules would be forming from whatever chemical precursors were present in the prebiotic 'soup'. At intermediate stages of prebiotic molecular synthesis, polymer synthesis is expected.

DISCUSSION

Motions of mica sheets are a possible source of energy for forming covalent bonds in prebiotic molecules, for changing the conformations of macromolecules, and for extruding lipid-enclosed protocells, as proposed recently, based on observations of L-form bacteria.[6] The spaces between mica sheets provide other advantages for the origins of life. When chemical reactions occur in confined spaces, there are fewer reaction products.[7] This would be an advantage, for example, for the formose reaction, which produces sugars by an autocatalytic process. Ribose is the only sugar currently found in RNA, but a great variety of sugars and branched sugars are produced by the formose reaction.[8] Perhaps the confinement between mica sheets, and the 0.5-nm-periodicity of the mica lattice, produce fewer sugars with the formose reaction.

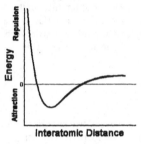

Figure 3. Mechanochemical synthesis of prebiotic covalent bonds is hypothesized to result from the close approach of molecules, into the attractive regime of the energy profile. Unlike thermal or chemical energy sources, mechanical energy can be uni-directional, thus favoring specific bond angles and reaction products. Orientation on the 0.5-nm mica lattice is also hypothesized to favor the polymerization of 0.5-nm-sized monomers such as the common amino acids and nucleotides.

Mica's chemical structure is similar to the chemical structures of many clays. Some of the advantages hypothesized here for mica apply also to clays. Others apply only to mica. The specific advantages of mica include the possibility of mechanochemistry with orders-of-magnitude variations in forces and distances, and the large closely spaced isolated compartments formed by the spaces between mica sheets. These advantages both stem from mica's property of having extremely large mineral sheets relative to clays.

Mechanochemistry and Nanomechanics

'Mechanochemistry' is the formation of covalent bonds by the use of mechanical energy.[9] In the origins of prebiotic life, covalent bond formation produced the monomers needed for biopolymers and the polymers resulting from polymerization of the monomers. These two synthetic processes would involve mechanochemistry on the picometer scale for synthesizing monomers and on the nanometer scale for synthesizing polymers.

'Nanomechanics' is used here to describe noncovalent rearrangements of biopolymers. These changes include molecular-level processes such as the aggregation/denaturation vs. renaturation of proteins, and cellular-level processes such as the blebbing off of lipid vesicles and their internal contents.

Mechanochemistry brings a directional component to reaction processes, unlike thermochemistry, in which the heat-induced forces are in random directions. The observations for mechanochemistry are that tensile stress makes reactions faster if the reactive site is elongated during the reaction and slower if the reactive site is shortened during the reaction.[9]

The energy of a carbon-carbon bond at room temperature is comparable to a mechanical force of 6 nN moving a distance of 100 picometers. Force can lower the energy barrier for chemical reactions, as shown by atomic force microscopy (AFM) and other single-molecule force-exerting techniques. Rupture forces, from AFM pulling measurements, are 1-2 nN for covalent bonds[10] and ~ 0.1-0.3 nN for the unfolding of protein domains.[11] Often, protein domains unfold and refold repeatedly as a protein with tandem domains is repeatedly stretched and relaxed.

Mechanical forces play a large role in living systems even at the molecular level. This is becoming increasingly obvious as the motions of single biomacromolecules are being explored. Fluid-filled spaces between mica sheets, moving on the nanoscale, have a resemblance to the nanoscale motions in subcellular structures such as enzymes and ribosomes. Hinge-like up-and-down motions are one of the most common movements in enzymes. These up-and-down motions might have arisen as an artifact of the up-and-down motions of mica sheets as shown in Figure 2. Other intra-macromolecular motions such as twisting, rotating (e.g., F1 ATPase), and walking (e.g., molecular motors such as kinesins and dyneins) are hard to relate to mica sheets' motions; and, of course, they probably evolved long after the Last Common Ancestor (LCA).

Intermolecular and surface forces of many types would have been involved in the hypothetical prebiotic mechanochemistries and nanomechanics. Molecules and surfaces interact within and between themselves through a multitude of forces, including electrostatic, hydrophilic, hydrophobic, van der Waals, solvation/hydration, protrusion, steric, and fluctuation forces.[12] Some of these interactions are sketched in Figure 1: One water-filled compartment shows tethered string-like molecules stretched between two mica surfaces. The next water-filled compartment shows bulbous molecules that will experience protrusion forces, and the bottom compartment shows molecules with good adhesion to mica. Intermolecular and surface forces act over distances ranging from less than 1 nm to over 100 nm.

Although research on mechanochemistry usually seems to result in bonds being broken, there are a few instances of synthetic mechanochemistry. For example, mechanically induced intramolecular rearrangements produced molecular products not formed with thermal or light-induced reactions.[13] Pressure is reported to induce the polymerization of glycine and to inhibit the degradation of poly-glycine.[14] Pressure affects the rate and equilibrium constants of chemical reactions.[15] In a destructive process, Molecular Dynamics calculations of thiols on copper showed that carbon-sulfur bonds cleaved when heat was the energy source, while copper-copper bonds cleaved when the energy source was an upward-moving mechanical force. Therefore the relative bond strengths were significantly different when the energy source was thermal as opposed to mechanical.[16]

Nanomechanics research, like mechanochemistry research, seems to show both breaking down and building up of molecular aggregates. For example: Pressure induces the aggregation of an amyloidogenic protein[17] and increases the aggregation of a IFN-gamma protein by exposing more of the protein surface area to solvent[18] On the other hand, pressure can renature and disaggregate some proteins. For example: Pressure causes aggregated human growth hormone to refold.[19] Pressure induces the non-covalent polymerization of G-actin monomers to F-actin fibrils in blood vessels.[20] Pressure can even induce the growth of blood capillaries.[21]

CONCLUSIONS

The origin of life is one of the major unanswered questions in science. Hypotheses for the origins of life, however, are hard to falsify convincingly and impossible to prove absolutely. Nonetheless, the mica hypothesis raises useful questions about the extent to which mechanochemistry can be used for molecular syntheses.

Thanks to: Robyn Hannigan, Arkadiusz Chworos, Robert Geller, Nathan Fay, and Megan Murphy for helpful discussions; Jim and Rich Greenwood, Connie Wieneke, and Joy Hansma for finding the mica mine that contributed to the origins of this hypothesis; and the National Science Foundation for funding the development and applications of atomic force microscopy, without which I would not have conceived this hypothesis.

REFERENCES

[1]   R. M. Hazen, *Genesis : the scientific quest for life's origin* (Joseph Henry Press, Washington, DC, 2005).

[2]   F. J. Dyson, *Origins of life* (Cambridge University Press, Cambridge [England] ; New York, 1999).

[3]   R. M. Pashley and J. N. Israelachvili, J. Colloid Interface Sci. **97**, 446 (1984).

[4]   H. G. Hansma and D. E. Laney, Biophys. J. **70**, 1933 (1996).

[5]   D. Schulze-Makuch and L. N. Irwin, *Life in the universe : expectations and constraints* (Springer-Verlag, Berlin ; Heidelberg, 2006).

[6]   M. Leaver, P. Dominguez-Cuevas, J. M. Coxhead, R. A. Daniel, and J. Errington, Nature **457**, 849 (2009).

[7]   P. Sozzani, R. W. Behling, F. C. Schilling, S. Briickner, E. Helfand, F. A. Bovey, and L. W. Jelinski, Macromolecules **22**, 3318 (1989).

[8]   G. L. Zubay, *Origins of life on the earth and in the cosmos* (Academic Press, San Diego, 2000).

[9]   M. K. Beyer and H. Clausen-Schaumann, Chemical Reviews **105**, 2921 (2005).

[10]   M. Grandbois, M. Beyer, M. Rief, H. Clausen-Schaumann, and H. E. Gaub, Science **283**, 1727 (1999).

[11]   R. B. Best, B. Li, A. Steward, V. Daggett, and J. Clarke, Biophys J **81**, 2344 (2001).

[12]   J. Israelachvili, *Intermolecular and Surface Forces* (Academic Press, New York, 1991).

[13]   C. R. Hickenboth, J. S.Moore, S. R. White, N. R. Sottos, J. Baudry, and S. R. Wilson, Nature **446**, 423 (2007).

[14]   S. Ohara, T. Kakegawa, and H. Nakazawa, Orig Life Evol Biosph **37**, 215 (2007).

[15]   S. W. Benson and J. A. Berson, Journal of the American Chemical Society **84**, 152 (1962).

[16]   M. Konopka, R. Turansky, J. Reichert, H. Fuchs, D. Marx, and I. Stich, Phys Rev Lett **100**, 115503 (2008).

[17]   Y. S. Kim, T. W. Randolph, F. J. Stevens, and J. F. Carpenter, J Biol Chem **277**, 27240 (2002).

[18]   J. N. Webb, S. D. Webb, J. L. Cleland, and J. F. Carpenter, PNAS **98**, 7259 (2001).

[19]   R. J. St.John, J. F. Carpenter, C. Balny, and T. W. Randolph, J. Biol. Chem. **276**, 46856 (2001).

[20]   M. J. Cipolla, N. I. Gokina, and G. Osol, FASEB J **16**, 72 (2002).

[21]   A. Mammoto, K. M. Connor, T. Mammoto, C. W. Yung, D. Huh, C. M. Aderman, G. Mostoslavsky, L. E. H. Smith, and D. E. Ingber, Nature **457**, 1103 (2009).

Mater. Res. Soc. Symp. Proc. Vol. 1185 © 2009 Materials Research Society   1185-II09-04

Elastic Properties of Nano–Thin Films by Use of Atomic Force Acoustic Microscopy

Malgorzata Kopycinska-Müller[1,2], Andre Striegler[1,2], Arnd Hürrich[3], Bernd Köhler[2], Norbert Meyendorf[2], and Klaus-Jürgen Wolter[1]
[1] Electronics Packaging Laboratory IAVT, Technical University Dresden, Helmholtzstr. 18 D-01069 Dresden
[2] Fraunhofer Institute for Nondestructive Testing IZFP-D, Maria Reiche Str. 2, D-01109 Dresden, Germany
[3] Fraunhofer Institute for Photonic Microsystems IPMS, Maria Reiche Str. 2, D-01109 Dresden, Germany

ABSTRACT

Atomic force acoustic microscopy (AFAM) is a non-destructive method able to determine the indentation modulus of a sample with high lateral and depth resolution. We used the AFAM technique to measure the indentation modulus of film-substrate systems M_{sam} and then to extract the value of the indentation modulus of the film M_f. The investigated samples were films of silicon oxide thermally grown on silicon single crystal substrates by use of dry and wet oxidation methods. The thickness of the samples ranged from 7 nm to 28 nm as measured by ellipsometry. Our results clearly show that the values of M_{sam} obtained for the film-substrate systems depended on the applied static load and the film thickness. The observed dependency was used to evaluate the indentation modulus of the film. The values obtained for M_f ranged from 77 GPa to 95 GPa and were in good agreement with values reported in the literature.

INTRODUCTION

Thin films play a fundamental role in microelectromechanical systems (MEMS), sensors and actuators, and electronics either as active functional coatings and membranes or passive protection layers. Rapid miniaturization that advances parallel to the development and use of new functional materials creates an increasing need for a non-destructive method that can be used for thin-film system characterization with high lateral and depth resolution. Mechanical properties of thin-film systems, such as indentation or Young's modulus of the film and the substrate, film thickness and its local variations, adhesion at the film-substrate interface, film density, and crystal structure and orientation are of importance for modeling and prediction of the film-substrate system behavior under different conditions.

Methods based on atomic force microscopy (AFM) [1] have been used to test mechanical properties of various materials and systems, including thin films [2,3]. There are different modes of AFM operation enabling one to measure Young's or indentation modulus of a sample, such as nanoindentation [4], interfacial force microscopy [5], and the so-called "acoustic" or "ultrasonic" methods [6-9]. The label comes from the characteristic frequency range (0.1 MHz– 3 MHz) in which the cantilever used vibrates. The acoustic methods offer the possibility of point measurements of the sample's elastic properties as well as elasticity related and/or calibrated images.

The AFAM technique, often called contact resonance spectroscopy method, has been successfully applied to characterize mechanical properties of thin-film systems, such as indentation modulus [10] and film-substrate adhesion [11]. In this method, an AFM rectangular cantilever is vibrated at ultrasonic frequencies while in contact with a sample surface. The elastic interaction between an AFM tip and the sample surface affects the resonance frequency of the vibrating system. The contact resonance frequencies are measured and used to calculate the effective tip-sample contact stiffness. Standard AFAM measurement procedures, experimental set-up, and theoretical models allowing for contact stiffness calculations are described elsewhere [12].

The AFAM technique is a contact based method, which means that a stress field is created under the AFM tip pressed into the sample at a certain static load. The size of the stress field depends on the applied static load, tip radius, and elastic properties of the tip and the sample. By a careful choice of the measurement parameters we can maintain a certain control over the size of the stress field, which ability creates a number of possibilities for new applications of the AFAM technique. One of them is the capability to detect subsurface defects, such as changes in the film – substrate adhesion and cavities [11, 13]. It has been also shown that it is possible to measure the elastic properties of 50 nm thin nickel films without sensing the elastic properties of the substrate [14].

In this study, we wanted to apply the AFAM method to determine the indentation modulus of very thin films, where the substrate influence would have to be separated from the data obtained. The AFAM measurements were performed on films of silicon oxide of thickness less than 30 nm. The results obtained for the indentation modulus of the silicon oxide were in good agreement with literature values. Details concerning the experimental procedure and data analysis process are presented in the following sections.

EXPERIMENTAL METHODS

The samples tested in this study were four films of silicon oxide thermally grown on silicon single crystal substrate. The thickness of the films ranged from 7 nm to 28 nm, as measured by ellipsometry. Parameters describing the sample preparation process and the resulting sample thicknesses are presented in Table I.

Table I. Parameters describing the silicon oxide thin films.

Sample #	Process	Temperature (°C)	Time (s)	Planned thickness (nm)	Measured thickness (nm)
1	"wet" oxidation	900	1500	28	28.6
2	"wet" oxidation	800	1380	16	16.2
3	"dry" oxidation	1130	30	10	11.4
4	"dry" oxidation	1130	10	5	7.6

As mentioned in the previous section, in AFAM experiments a stress field is created under the AFM tip. The size of the stress field defines the volume of the tested sample that provides the information on the local indentation modulus M_{sam}. In the AFAM experiments reported here, the experimental parameters were chosen such that for the lowest static load F applied to the tip the resulting stress field was contained mostly within the film. Figure 1 (a) presents the schematic of the idea behind the measurement. The curves plotted for the stress field σ(z) were estimated for an ideal hemispherical silicon tip of $R = 30$ nm contacting a silicon oxide film of $M_f = 80$ GPa and thickness $t = 15$ nm grown on a silicon single crystal substrate <100>-oriented. As the values of F increase, the stress field expands into the substrate, causing its deformation sufficient to dominate the deformation of the whole compressed volume. Figure 1 (b) shows example of contact stiffness values obtained for silicon oxide film 16 nm thin and two reference samples of fused quartz and silicon single crystal. As can be seen, the values of k^* obtained at the lowest static loads for the thin-film sample were very similar to those obtained for the fused quartz sample. As the static load increased, the values of k^* obtained for the thin-film increased rapidly. At the highest static loads used in these experiments, the values of k^* obtained for the thin-film sample were much greater than those obtained for the fused quartz but still not as large as the values of k^* obtained for the silicon sample.

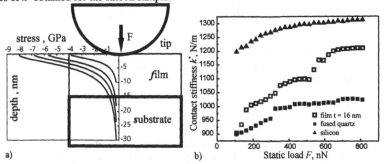

Figure 1. (a) Schematic representation of stress field expanding under an AFM silicon tip at increasing static loads. (b) Experimental values of local contact stiffness k^* obtained from AFAM measurements performed on silicon, fused quartz and thin-film sample.

A rectangular AFM beam with free resonance frequencies of 153 kHz and 938 kHz for the first and second mode, respectively, has been used to measure the contact resonance frequencies. The cantilever spring constant k_c was about 27 N/m as estimated from the cantilever dimensions and the value of the first free resonance frequency. The contact resonance frequencies were measured for the first and the second contact modes at static loads increasing from 30 nN to 800 nN in 30 steps. Reference measurements proceeded and followed each single measurement performed on one of the thin film samples. The thin-film samples were measured six to eight times each. The cantilever tip was characterized by use of two reference samples, namely fused quartz of $M_f = 77$ GPa and silicon single crystal $M_{Si<100>} = 165$ GPa. The choice of the reference samples was dictated by the similarity of their elastic properties to those of the films and the substrate.

The values obtained for the contact resonance frequencies were used to calculate the values of the effective tip-sample contact stiffness k^*. The values of k^* were then employed to calculate the values of the indentation modulus of the samples by use of Hertz model for contact mechanics [15]

$$k^* = 2aE^* = \sqrt[3]{6RFE^{*2}}$$ (1).

113

Here, a is the contact radius, R is the tip radius, F is the static load applied to the tip, and E^* is the reduced Young's modulus defined for a homogeneous sample as follows [16]:

$$\frac{1}{E^*} = \frac{1}{M_t} + \frac{1}{M_{sam}} ,$$ (2)

where M_t and M_{sam} are the indentation moduli of the tip and the sample, respectively. In case of the thin-film system one can define the elastic modulus of the sample in a way similar to that proposed by Swein et al. [17]:

$$\frac{1}{M_{sam}} = \frac{1}{M_s} + \left(\frac{1}{M_f} - \frac{1}{M_s} \right) e^{-a\frac{F}{t}} .$$ (3)

Here, M_s and M_f are the indentation moduli of the substrate and the film, respectively, t is the film thickness and α is a factor depending on the tip geometry and the ratio of the contact radius and the indentation depth [18].

RESULTS AND DISCUSSION

The values of the indentation modulus obtained for the film-substrate systems are presented in Fig. 2. They were obtained by use of a standard AFAM procedure described in detail elsewhere [19]. We used values of contact stiffness obtained for the fused quartz sample as the reference. The values of M_{sam} were first evaluated for each single measurement as a function of the static load and then averaged. The error bars plotted in Fig. 2 are the standard deviations of the M_{sam} mean value. According to our expectations, the values of M_{sam} obtained for the thin-film samples are nonlinearly dependent on the applied static load. In addition, the dependence of M_{sam} on F is different for each of the films, which allows us to distinguish between the samples, even if the thickness variation is only 4 nm. One can also see that the mean values of M_{sam} calculated for the 7 nm thin film have the largest error bars. In case of the thinnest film we may have observed the largest influence of the substrate topography and thickness variations. In addition, we observed sudden increases in the values of M_{sam}, which can be only explained by fracture and delaminating of the film under the tip. Such occurrences took place for all of the tested samples, however, they were most numerous for the 7 nm thin film. As we learned from experience, use of a tip with a smaller tip radius and smaller static loads allows avoiding such events and improves the accuracy of the measurement.

The dependence of M_{sam} on F has been used to evaluate the indentation modulus of the film M_f. We used eq. (3) and the experimental dependence $M_{sam}(F)$ to find the values M_f for each of the tested films. We used the values of the $M_{Si<100>} = 165$ GPa and the film thicknesses t as fixed parameters. The parameters M_f and α were obtained from the best fit curve. Figure 3 shows plots of the data points obtained for the 16 nm thin-film sample and the best fit of the theoretical model. The fit procedure was performed for each individual $M_{sam}(F)$ curve. Then the results obtained for M_f were averaged. The mean values obtained for the indentation modulus of the film and the corresponding standard deviation values are presented in Table II.

114

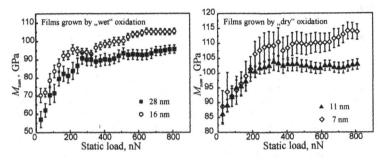

Figure 2. Results obtained for the indentation modulus M_{sam} of thin-film samples of different thicknesses.

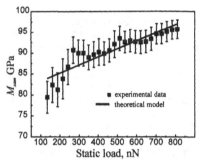

Figure 3. Comparison of the experimental values of the indentation modulus obtained for the 16 nm thin-film sample and the values calculated from the best data fit parameters.

Table II. Mean values of the thin film indentation modulus M_f and corresponding standard deviation.

t (nm)	28	16	11	7
M_f (GPa)	77.5 ± 8.7	88.1 ± 10.2	94.3 ± 8.9	94.9 ± 12.1

The values obtained for the indentation modulus of the silicon oxide films range from 77 GPa to 95 GPa. They are very similar within the error bars and in a very good agreement with values reported in the literature [20]. However, one notices the increase in the values of the indentation modulus M_f as the film thickness decreases. There are two plausible theories that may explain this behavior. The first one says that the thicknesses of the two thinnest films were comparable to the contact radius and thus the results obtained for the two samples could not be analyzed with our simplified model. The second theory takes into account that the 7 nm and 11 nm films were grown at higher temperatures than the 16 nm and 28 nm films. Thus we may have observed the influence of the sample preparation process on the sample elastic properties.

SUMMARY

We used the AFAM technique to study the elastic properties of thin-film samples. The thickness of the silicon oxide films ranged from 7 nm to 28 nm. In the first step, we determined the values of the indentation modulus of the film-substrate system as a function of the applied static load and observed a nonlinear dependence of $M_{sam}(F)$ for all of the tested samples. In addition the function $M_{sam}(F)$ was different for each of the samples. We could use the obtained results to identify the samples, even if the difference in the film thickness was only 4 nm. As a next step we used the dependence $M_{sam}(F)$ to determine the indentation modulus of the film for each of the tested thin-film samples. The values obtained for the M_f ranged from 77 GPa to 95 GPa and were in a good agreement with the literature values reported for Young's and indentation modulus of silicon oxide films.

LITERATURE

1. G. Binning and C. F. Quate, *Phys. Rev. Lett.* **56**, p.930 (1986).
2. N. A. Burnham, R. J. Colton, *J. Vac. Sci. Technol. A* **7**, p. 2906 (1989).
3. E. K. Dimitriadis, F. Horkay, J. Maresca, B. Kachar, and R. S. Chadwick, *Biophys. J.* **82**, p. 2798 (2002).
4. D. Tranchida, S. Piccarolo, *Macromol. Rapid. Commun.* **26**, p. 1800 (2005).
5. S. A. Joyce and J. E. Houston, *Rev. Sci. Instrum.* **62**, p. 710 (1991).
6. R. E. Geer, O. V. Kolosov, G. A. D. Briggs, and G. S. Shekhawat, *J. Appl. Phys.* **81**, p. 4549 (2002).
7. M. T. Cuberes, H. E. Assender, G. A. D. Briggs, and O. V. Kolosov, *J. Phys. D. Appl. Phys.* **33**, p. 2347 (2000).
8. K. Yamanaka, Y. Maruyama, T. Tsuji, and K. Nakamoto, *Appl. Phys. Lett.* **78**, p. 1939 (2001).
9. U. Rabe, K. Janser, and W. Arnold, *Rev. Sci. Instrum.* **67**, p. 3281 (1996).
10. D. C. Hurley, K. Shen, N. M. Jennett, and J. A. Turner, *J. Appl. Phys.* **94**, p. 2347 (2003).
11. D. C. Hurley, M. Kopycinska-Müller, E. D. Langois, A. B. Kos, and N. Barbossa III, *Appl. Phys. Lett.* **89**, p. 021911 (2006).
12. U. Rabe, S. Amelio, E. Kester, V. Scherer, S. Hirsekorn, and W. Arnold, *Ultrasonics* **38**, p. 430 (2000).
13. Z. Parlak and F. L. Degertekin, *J. Appl. Phys.* **103**, p. 114910 (2008).
14. M. Kopycinska-Müller, R.H. Geiss, J. Müller, and D.C. Hurley, *Nanotechnology*, **16**, p. 703, (2005).
15. K. L. Johnson, *Contact mechanics*, (Cambridge University Press, Cambridge UK, 1985), p. 90-96.
16. J. J. Vlassak and W. D. Nix, *Phil. Mag. A* **67**, p. 1045 (1993).
17. M. V. Swein and E. R. Weppelmann, *Mat. Res. Soc. Symp. Proc.* **308**, 177, (1993).
18. R. B. King, *Int. J. Solids Structures* **23**, p. 1657 (1987).
19. U. Rabe, M. Kopycinska, S. Hirsekorn, J.M. Saldana, G.A. Schneider, and W. Arnold, *J. Phys. D-Appl. Phys.* **35**, p. 2621 (2002).
20. H. Ni, X. Li, and H. Gao, *Appl. Phys. Lett.* **88**, p. 043108-1 (2006).

Printed in the United States
By Bookmasters